Airplane
Ownership

Airplane Ownership

Second Edition

Ronald J. Wanttaja

McGraw-Hill

New York   Chicago   San Francisco   Lisbon   London   Madrid
Mexico City   Milan   New Delhi   San Juan   Seoul
Singapore   Sydney   Toronto

The McGraw·Hill Companies

Library of Congress Cataloging-in-Publication Data

Wanttaja, Ron.
   Airplane ownership / Ron Wanttaja.—2nd ed.
      p.   cm.
   Rev. ed. of: Airplane ownership. c1995.
   Includes index.
   ISBN 0-07-145974-X
   1. Private planes—Purchasing. 2. Private planes—Maintenance and repair. I. Wanttaja,
Ron. Airplane ownership. II. Title.

TL671.85.W36   2006
629.133′34022—dc22   2005053521

3 4 5 6 7 8 9 0 DOC/DOC 0 10 9 8 7

ISBN 0-07-145974-X

The sponsoring editor for this book was Steve Chapman and the production supervisor was Richard C. Ruzycka. It was set in Slimbach Std. by International Typesetting and Composition. The art director for the cover was Anthony Landi.

Printed and bound by RR Donnelley.

McGraw-Hill books are available at special quantity discounts to use as premiums and sales promotions, or for use in corporate training programs. For more information, please write to the Director of Special Sales, McGraw-Hill Professional, Two Penn Plaza, New York, NY 10121-2298. Or contact your local bookstore.

Contents

Introduction

AMONG THE NON-FLYING PUBLIC, aircraft ownership is a sign of the well-to-do.

It's definitely a deeply-imbedded belief. I once told a ten-year-old neighbor that I'd take him for a ride in my plane.

His response: "Are you rich?"

It's one thing when the non-flying public believes you have to have bottomless pockets to own your own plane. Too many pilots have the same impression.

Let's not kid ourselves. Planes aren't *cheap*. You'll never own a plane for the same cost as a well-used car. But just because they cost more than a dented Kia doesn't mean you can't afford one.

Aircraft ownership—or access equivalent to ownership, such as a partnership—is available to just about anyone who's willing to work at it.

In this book, we'll explore the costs, benefits, and responsibilities of aircraft ownership. We'll take a look at whether aircraft ownership is for you. We'll explore the process by which you can find the best airplane for the money. We'll examine the ways you, as a new aircraft owner, can minimize your expenses. And, finally, we'll look at how to minimize the effect of problems that occur along the way.

Ronald J. Wanttaja

Acknowledgments

IT TAKES A VILLAGE TO RAISE A CHILD, but when the village idiot decides to write a book, a lot of help is needed.

Thank you to my case study participants, for their candor, willingness to help, and the photos of their aircraft that grace this book. Thanks to study subjects Eric Warren, Lynn Berkell, Kathleen and Patrick Flynn, Miles Erickson, Don Connell and "Stearman Stan" Brown, Mike Furlong, Rich Shankland, "Mid Life Crisis Aviation," Mary and Jay Honeck, and Javier Henderson. Thanks also to Tim Elliot, for the use of his photographs.

Thanks to the gang at Auburn (Washington) Municipal Airport, for their willingness to open their hangars to my camera, and their tolerance of a bunch of darn-fool questions.

And once again, thanks to Lisa, my bride of twenty-five years. She still provides what every writer needs the most: Good editing and chocolate chip cookies.

Airplane
Ownership

1

Why Buy?

It's great being a pilot, isn't it?

You can come up with a thousand-and-one reasons. The freedom. The vast distances you can quickly travel. The separation from the everyday world, in more than just altitude. But too often, there's a question that'll bring you back to earth.

You'll be talking with some new friends, and it'll somehow come out that you're a pilot. You'll extol the joys of flying, maybe even tell a funny story from your student days.

Then someone asks: "Do you own a plane?"

When you tell them, "No," there's always an odd little pause. "I rent," you say weakly, to fill the silence.

You can imagine what they're thinking. You've just explained how flying is no harder than driving a car—now you're admitting you have to borrow an airplane for every little flight. Imagine having to beg the use of a car just to go to the store for a loaf of bread!

RENTING IS THE PITS

Be honest. Do you *like* renting? Surely, a few folks enjoy flying a variety of aircraft and prefer not to be tied down. But face it, renting is the pits. You can't fly when you want. How often have you asked, "Is anything available *Sunday* afternoon, then?"

Do you have the hankering to fly something other than plain-vanilla 172s and Pipers? There may be some fun airplanes, like Aeronca Champs (Fig. 1-1), available for rent in your area, but their rental rates are often disproportionately high.

Have you ever had to cut short an enjoyable afternoon jaunt to get the plane back to the airport in time for the next renter? Or rush to get home before the Fixed Base Operator (FBO) closes (Fig. 1-2)?

Surely you'd rather own your own airplane. Why don't you? Two reasons come to mind. The first is *cost*. Everyone knows airplanes are expensive. Especially the nonflying public. We're all "Fat Cats," according to Ann Landers. If your neighbor

Fig. 1-1. *Classic aircraft like this Aeronca Champ aren't usually available on the rental market.*

buys a $50,000 sports car, everyone admires him. If you buy a $30,000 airplane, everyone frowns and mutters that you have too much money for your own good. Yet your neighbor might pay more every year on his sports car's insurance premium than some planes cost to operate for a year—including quite a bit of flying!

The second reason you don't own an airplane is related: Fear of the unknown. You'd like to know what you're getting into before taking the big step. What'll it *really* cost? How are you going to handle a sudden maintenance bite? What do you do if you have to force-land in a farmer's field?

I can't help with your neighbors. But I can reduce your anxieties about airplane ownership. That's what this book is about. I'm going to give you the information to decide if ownership is for you. We'll discuss the costs involved and ways to reduce them.

Fig. 1-2. *If you owned a plane, you could keep it out as late as you wanted. (Courtesy Cessna Aircraft Company.)*

You'll see the whole buying process, and the pitfalls which may lurk beneath the rosiest deal. We'll look at maintenance procedures the owner may perform and how to do them safely. We'll grit our teeth and examine possible actions if ownership turns sour.

In short, we're going to look at owning an airplane from the real-world perspective. By the time you finish this book, you should have a good idea as to whether you can handle—and afford—owning your own plane.

THE SORDID TRUTH ABOUT OWNERSHIP

To begin with, I'm going to tell you something really ugly: In all likelihood, for the amount of flying you're probably going to do, it's cheaper to rent than own.

Take a look at Fig. 1-3. It's a computation for the cost per flight hour for a used Cessna 172 for a variety of annual utilization rates. The shaded area shows how the per-hour cost varies depending upon the ownership situation—the lower bound is a low-cost ownership situation and the upper bound represents a more expensive ownership case. The dark line down the center is the average of these two cases. The horizontal line shows the typical rental rate for Cessna 172s. Note how the average curve doesn't cross the rental rate line unless the owned plane is being flown at least 140 hours per year.

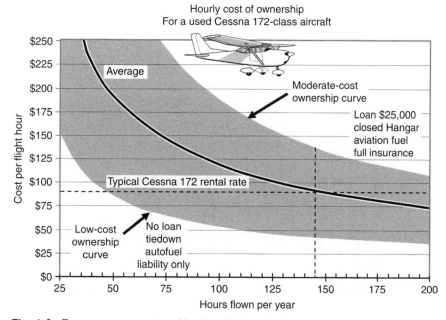

Fig. 1-3. *For an average ownership situation, an airplane must be flown at least 140 hours per year to get a per-hour cost that's less than renting the same plane.*

That's a lot of flying time. Almost 3 hours a week—each and every week. The average General Aviation pilot flies about a 100 hours per year. So why buy? Why not just keep renting?

The key to owning

The key to justifying ownership over renting doesn't lie in comparing the per-hour rates. The question is not, "Is it cheaper to rent?" The question is: "Can you afford to own?"

In other words, stop looking at justifying ownership with the prospect of saving money over renting. Instead, decide that you have a new *hobby*: Airplanes. Compare it to another hobby—stamp collecting, for instance. How many philatelists' activities are self-supporting? Few, I suspect. There are undoubtedly some wheeler-dealer types who turn a profit. But most are content to spend a dollar here and a sawbuck there to give them a little pleasure.

Why should owning a Tobago (Fig. 1-4) be any different?

Unfortunately, unless you're collecting those funny stamps with the airplanes printed upside-down, your flying hobby is going to cost a bunch more than stamp collecting. Hobbyists have to decide how much of their income they can afford to allocate to their favorite pursuit. I don't know how many aircraft are operated by "hobby" owners, but the percentage is high; really high.

It boils down to how much you can afford. Or what you're willing to give up. Most owners make some sort of sacrifice. Maybe they forgo buying a new car. Or live in an older house or even a small apartment. They vacation in the pilot's seat, instead of Europe. Just so the money is there to keep the plane flying.

In Chap. 2 we'll dissect ownership costs to help you determine if there's an ownership plan that'll fall within your budget. But for right now, you're trying to justify owning. Let's look at some of the advantages.

Fig. 1-4. *Consider airplane ownership a hobby, and not as a way to escape FBO rental fees. (Courtesy Aerospatiale General Aviation.)*

OWNERSHIP ADVANTAGES

The biggest advantage to owning your own plane is that it's there when you want to fly. Its schedule is entirely subservient to your whims.

Say you wake up to a gorgeous Saturday morning. Beautiful day to fly. But you don't own a plane. You dive for the phone and call your local FBO. Rats. Everything's taken.

If you owned, you'd just drive the airport. Your plane's ready to preflight and go.

And while you're flying, you can go where you wish and return when you feel like it. No one's sitting in an office waiting for you to bring their plane back.

If you want to use an airplane for transportation, owning is even better. Let's say you want to fly to visit a friend 300 miles away, stay two nights, and return.

If you rent, it'll be necessary to schedule the airplane well in advance, especially if the trip is on a weekend. Two weeks or more, at some FBOs.

And what if the weather's bad the first day? You'll have to either shorten the trip or lose your chance. The plane's probably booked on the day following your scheduled return, so you can't just slide the trip back.

If you owned, you could leave when you wanted.

Back to the rental. To get a solid 3-day block of time, you usually have to pay a certain minimum. It's usually equivalent to around 3 hours of flight time per day.

So although you'll probably make the round trip in less than six total flying hours, you'll pay for nine. The effective per-hour rate just rose by 50 percent.

And if the weather closes in while you've got the plane at your friends'? The FBO keeps charging 3 hours' worth every day until the plane is returned. And if you have to leave it there to return home? You'll pay for the planes and pilots sent to retrieve your abandoned rental.

Sure, it wouldn't be a picnic if you owned the aircraft. You'd have to arrange to tie the plane down at the destination airport until you can return to collect it. Still, a couple of weeks' tiedown fee should be well less than $100.

Another advantage of owning has to do with your own skills. You'll fly more often if you own; a current pilot is a safer pilot. You'll become familiar with the handling and quirks of your own plane. Should an emergency situation arise, you'll be more capable of handling it. I once had a fuel-contamination engine failure in my old 150, and managed to milk the glide well enough to make a deadstick landing at a nearby airport.

Could I have done as well in a rental airplane? I'd like to think so. But in the preceding weeks, I'd gotten interested in glide-path control. I'd practiced the techniques in Langewiesche's Stick and Rudder. The practice undoubtedly contributed to the successful outcome, whether I'd been flying my own plane or not. However, I believe that my intimate familiarity with the glide characteristics of Cessna Zero-Niner Tango pushed the odds significantly in my favor. All airplanes fly differently, no matter what my fellow engineers say.

Another factor sometimes enters into the formula: mechanical reliability. Some owners are mistrustful of the degree and quality of maintenance rental aircraft receive, and feel their own planes are more reliable due to their personal attention. Perhaps.

There are FBOs and clubs that scrimp and penny-pinch on maintenance, and there are those that take pride in the reliability of their fleet. Similarly, there are owners who spare no expense when it comes to maintaining their aluminum steed, and there are those who spend just enough to get their plane to pass each annual inspection.

Which are you?

In either case, at least your plane won't be subjected to the rough usage that rental aircraft receive. Better reliability results. You'll have more confidence in the aircraft since you'll be familiar with its recent flight and maintenance history. You'll also be able to optimize the plane to your own needs and desires. Do you require extra cushions to sit comfortably? You can leave 'em in the airplane, rather than dragging them back and forth to the airport. Like to fly from shorter airports? Have the prop repitched to reduce take-off distance. Want nose art to personalize your machine? Get out the brushes or head to the sign shop (Fig. 1-5).

There are other, less tangible benefits to ownership. It's a good excuse to hang around the airport all day. Watching the touch-and-go traffic while sitting under the wing of your own plane is a marvelous way of killing a Saturday afternoon. Ownership is the entre' to a fellowship, which includes some of the finest people in the world. A couple of hours spent changing oil and waxing your bird will see a number of them come by to exchange stories and advice.

Fig. 1-5. *You own the airplane—why not add insignia or markings to personalize it?*

Finally, there's often one financial advantage to ownership: If it's the typical used General Aviation bird, if you keep the plane maintained and undamaged, it'll probably sell for more than you paid for it. The average General Aviation airplane is about 27 years old; the prices on most planes of this vintage or older have bottomed-out. Some planes with classic appeal even make pretty decent investments.

OWNERSHIP DISADVANTAGES

If it breaks, it's going to cost big money to fix it.

That's probably your biggest worry. It's the fear of the sudden maintenance bill that scares away most potential owners. It can cost $25,000 or more to overhaul some aircraft engines (Fig. 1-6). Scraping together that much money is tough for most people, especially if they've already gone into hock to buy the plane in the first place.

It is a possibility. We can reduce the chance of major maintenance expenses by a careful prepurchase inspection. We might be able to anticipate and prevent major failures with a thorough maintenance plan. And there are ways to mitigate and stretch out the costs, should the worst occur. We'll discuss them later.

But they *can* happen.

Another disadvantage to ownership is that it essentially restricts you to flying only one airplane. If you own, it's hard to justify the additional expense of renting a plane just for a change of pace; Especially to a spouse. And, depending upon the aforementioned spouse, it can be a source of family tension. Your mate may not be as fond of flying as you are. An airplane is a hole in the sky, into which you throw money; this can produce understandable resentment. Most wives and husbands can come up with excellent alternative uses for the $1000 per month your winged chariot might cost.

Fig. 1-6. *The unexpected need for an engine overhaul is probably the biggest financial risk in airplane ownership.*

Similarly, if the plane takes up too much of your monthly income, it dulls the joy of ownership. The solution lies in making the most accurate determination of costs prior to purchase and in a realistic assessment of how much you can budget toward ownership.

This book should help. Chapter 2 discusses the financial aspects of ownership, and options which can reduce costs. Chapter 3 helps you pick the airplane that fits your needs, Chaps. 4 and 5 discuss the advantages and disadvantages of new and used airplane, and Chap. 7 guides you through the purchase process.

THE FAMILY ELEMENT

Oh, oh.

We're going to talk about how your spouse is going to look at your decision to buy an airplane. Maybe you're a carefree bachelor and don't need to justify your spending. Maybe you make so much money that an airplane won't even dent the family budget.

But it probably isn't so. There's a running joke among the homebuilder clan. Building your own plane requires enormous dedication, often to the exclusion of family relationships. Anyway, if you ask a homebuilder how much his or her plane cost, you'll sometimes get answers like:

"$20,000 and Brenda," or, "$30,000, not including alimony."

You can get a similar response even when you just *buy* a plane. It's natural for your wife or husband to view the "new member of the family" with worry.

My old 150 was a case-in-point. I bought it just before we moved out of a trailer into a real house. Our new living room stood empty of furniture during the two years I owned that plane. My wife spoke wistfully of someday filling that empty space. But bless her heart, she never challenged the aircraft as the obstacle to her dream. The day after I sold the plane, though, she went out and spend half the money on new furniture.

How can you prevent or minimize family friction over your new "toy?" I'm no marriage counselor. But I'm a firm believer in telling the truth up front.

So unless the numbers really, seriously, show it, don't justify it by claiming owning will be cheaper than renting. Take my word for it, it probably won't. Even if your numbers show an ownership edge, a mechanical problem can negate all your careful figuring. And if you've bought the airplane "cause it'll save us money," you'll be in deep doo-doo.

Another way of forestalling problems is to buy the aircraft that meets your spouse's level of interest in your new hobby. Say, for instance, you convince him or her that the plane will be useful for long family trips. If that's the case, *don't* buy a small two-seater! Get something with a little room—something with luggage space. And if it doesn't have room for Junior, you'll be nose-deep in that smelly stuff again.

On the other hand, if your spouse isn't interested in participating, don't go hog-wild with the aircraft selection. You won't need a Cirrus (Fig. 1-7) for zipping out for a hamburger on Saturday afternoon.

Fig. 1-7. *Planes like this Cirrus SR-20 are excellent for fast, comfortable, long-distance travel, but a simpler airplane may be better suited for your family situation. (Courtesy Cirrus Design Corporation.)*

Speaking of Saturday afternoons, remember that your spouse may actually be more jealous of your time than of the money spent (Fig. 1-8). Homebuilders know this problem well—they spend hours in the shop, showing up at mealtimes as a bleary-eyed mound of sawdust. Though homebuilders' mates have problems you won't encounter. A wife may tolerate scrapes from razor stubble, but most draw the line at epoxy dermatitis.

Disappearing to the airport every weekend can cause friction. Keep that in mind, lest you end up with no home to go home to. Tell the truth, and remain attentive and sensitive to your mate's needs; a good prescription for any marriage, with or without airplanes.

Fig. 1-8. *The owner of this RV-6 homebuilt recognizes how his wife might see his airplane.*

ANSWERS TO COMMON QUESTIONS

Before we move on, there are several questions many prospective owners have. Let's get them out of the way.

Should I buy a plane to learn to fly in?

If you're a student, or haven't started lessons yet, buying your own plane to take lessons in can seem an attractive option. After all, the per-hour rate decreases the more you fly. Availability is never an issue. You'll be able to fly more often and practice whenever you want.

However, I don't recommend this course. Why?

1. Training is hard on an airplane. A bad landing hurts bad enough with the instructor sitting in the right seat of a rental bird. Why put that wear and tear on your own machine? Sure, if you buy a used 150 or 172, it probably has thousands of hours in the sweaty grip of student pilots. But why add to it?

2. Insurance is going to cost more. If you're just carrying liability, the hit may not be as bad. But if you carry hull coverage (insurance to cover accident damage), the rates will be higher for a student with 10 hours vs. a private pilot with a hundred.

3. It may limit your options for financing. If the plane will be used as collateral on the loan, the bank may be more leery of lending the money to a student pilot.

4. Aircraft availability may restrict your training. If you're renting and your selected aircraft becomes unserviceable, there's usually another one to choose. Not so if your own plane requires work. If a part ends up backordered, your lessons may screech to a halt for weeks.

5. Picking the right plane is difficult. Did you buy a car before you learned to drive? If you don't have a number of flying hours already, you won't be able to tell bad characteristics from good. Sure, you can ask a flying buddy or your instructor to fly your prospective purchase. But wouldn't you really rather decide for yourself? If nothing else, wait until you've got 30–40 hours under your belt. You'll then have enough background to make an informed choice.

6. A good trainer may not be the right airplane for long-term ownership. Cessna 150s are good planes to learn to fly in, but they're rather slow and cramped. If you intend to fly a lot of long cross-countries after you get your license, a Warrior or 172 might be a better pick.

Note that these drawbacks are mostly intangible; it's more a collection of "What ifs" than definite iron-clad problems you'll run into. I've known people who have bought planes to take lessons on, and everything turned out OK.

One case where I do feel buying a plane for instruction might work out better is if you're going for an advanced rating. If you've already got your private and are working on your instrument or commercial, numbers 1 through 3 discussed earlier don't really apply. Number 6 actually works in your favor, as the larger airplanes generally are better equipped and are better instrument platforms.

Can I save money by buying a plane with a bad engine and installing a converted automobile engine?

This is something I thought about prior to buying my first plane. The engine is the most expensive part of an airplane. If the engine is ruined, the airframe itself can sometimes be picked up at an extremely low price. The homebuilt magazines are full of companies selling Chevrolet, Subaru, Ford, Honda, and other engines converted for aircraft use. Why not put one of these engines in a Cessna or Piper, and put it in the "experimental" category like a homebuilt?

The simple answer is: You can't.

Oh, there are legal ways of doing this. But none that will work for the average aircraft owner.

The "experimental" category isn't just homebuilts. There are a number of subgroups beneath it, all with different requirements and limitations. Homebuilts fall into the "amateur-built" group. Amateur-built experimental aircraft have the least amount of limitations placed on their operations. They are subject to tight restrictions during a flight-test period, typically 25–40 hours. After that, they can't be flown over congested areas (except flying to and from airports) and can't be operated for hire (i.e., rented out or otherwise used commercially).

There is one big requirement that must be met before a plane can be certified as experimental/amateur-built: At least half the airplane must be constructed by the builder.

Obviously, a Cherokee with a Ford engine wouldn't qualify. The Federal Aviation Administration (FAA) allows such an airplane to be placed into the experimental/Research and Development (R&D) or experimental/market survey categories.

The R&D category is for testing new aviation concepts, from airfoils to electronics to, yes, engines. Once the modification has proven to be viable, aircraft can move into the market survey category to allow the developers to "test the waters" with regard to placing their new concept in production. Even if you did get the plane into these categories, the FAA usually restricts its operations. You might not be authorized to carry passengers, or any flight farther than 100 miles might require prior approval.

The 172 shown in Fig. 1-9 is powered by a Chevrolet V-6 engine. A Cessna airframe was the most expeditious way to get the powerplant aloft. This 172 is certified in the experimental/market survey category to allow the owner to take it to airshows across the country to help sell engines to homebuilders.

If you develop an auto-engine conversion, the FAA is quite willing to give you the regulatory permission necessary to fly the aircraft. If all you want to do is install someone else's converted engine, there is no justification for allowing the aircraft into either the R&D or market survey categories.

Fig. 1-9. *A Cessna 172 powered by a Chevrolet V-6 engine. The FAA allows persons or companies developing auto-engine conversions to use certified aircraft as test beds, but restrictions limit their usability as personal aircraft.*

That's not to say you can't make any changes to your aircraft. But modifications are restricted to those, which include approved components. We'll talk about the modification process in Chap. 11.

How often does a plane require mechanical inspection?

While renting, you may have noticed that planes occasionally go out of service for two types of inspections: "annuals" and "100 hours."

In most cases, a privately-owned aircraft only requires an annual inspection; that is, once a year. The 100 hour only applies to those aircraft operated "for hire." So, the aircraft provided by your FBO requires a 100 hour, but your own plane won't.

Unless, of course, it's on "leaseback" to the FBO. Leaseback is when you lease your plane to a commercial operator for rental or charter. In these cases, the 100 hour is required, and you (as the owner) will have to pay for it.

Should I select a plane based on fuel economy?

It really depends on the current price of fuel. Traditionally, the direct costs of airplane ownership (i.e., fuel and oil) were overpowered by the fixed costs: hangar, insurance, maintenance, and the like. While a plane that burned less fuel will (naturally) cost less to run, the difference was often too trivial to use it as a selection criteria.

However, skyrocketing fuel costs in the mid-2000s have upset the traditional relationships. It's best to study how the cost of ownership will vary with fuel efficiency.

Can I use the plane in my business?

Of course. As long as your business isn't transporting people or cargo, you can use the plane to support it however you wish. If a remote site needs a part, you can fly it there. If you've got a business meeting in another town, you can take your own plane.

When I was in a small club, one of our members was an insurance adjuster. He delivered claim checks in our little open-cockpit homebuilt. He didn't need a commercial pilot license, since the aircraft use is incidental to his employment.

Where the FAA draws the line is, if you start acting like an airline or charter service. As long as you don't "Hold out as" (their term) a transportation service, you should be OK.

You can also use the airplane to commute to your work. I've known several folks who have done this, with various levels of success. One says, "Choose a commuting arrangement that has a ground-based backup, and don't so heavily favor the airplane commute that one is tempted to force the issue and go flying when it's dangerous to do so." He used a Cessna 182, and in over three years of commuting, had only about $1000 in additional maintenance costs.

A NOTE ON THE CASE STUDIES

Throughout this book, you'll be meeting airplane owners in a number of case studies. You'll see the kinds of problems and the kinds of successes folks in a wide variety of ownership situations have seen. You'll meet the proud owner of a brand-new Cirrus. A transportation planner who takes long trips in a humble Cessna 150, and a woman with a classic Aeronca. You'll see examples of leaseback and partnership situations.

And—I guess it's only fair to start with myself!

CASE STUDY: FROM MINIMUM TO MODERATE

A small number of key decisions make a tremendous difference as to the total cost of airplane ownership. Comparing the impacts of these decisions is normally quite difficult. Too many variables change—the airplane types are different, the owners are different, the planes are based at separate airports, and so on.

But our first case study is different. We can compare the cost of ownership between two examples of *exactly* the same type of airplane. Based at *exactly* the same airport. And operated by *exactly* the same guy—me.

The differences? The type of ownership, the level of aircraft equipment, and the general philosophy involved in keeping the airplane. One was operated in the early 90s as a club project, and the other is my current personal pampered pet. Both aircraft were examples of the Bowers Fly Baby, a homebuilt designed in the early 60s by the late aviation writer—Peter M. Bowers. The design won the first Experimental Aircraft Association (EAA) design contest in 1962, and Mr. Bowers sold more than 5000 sets of plans. Over 400 examples have been constructed by amateur builders.

Fig. 1-10. *The author's current aircraft, a 1982 Bowers Fly Baby homebuilt.*

Our club's airplane, in fact, was N500F, Bowers' original prototype. We operated it from 1986 through 1994. In 1996, I decided to get back in to Fly Baby flying, and bought my own (Fig. 1-10).

Let's compare the two situations. To give a level playing field, the expenses of the older airplane have been adjusted to what they would be today.

N500F: As cheap as it gets

I used to get a lot of people mad at me. Sometimes when pilots get together, we talk about how much our airplanes rob our pocketbooks. I usually sat back and listened to the sob stories. Then I chimed in with my own numbers: N500F cost about $1600 per year to own (adjusted for inflation). That's about $140 per month. Divided between the three members of the club, that was less than the monthly fees for deluxe cable TV.

That got their attention.

Our operating expenses were low, as well: The direct costs ran about $18 per hour, including fuel, oil, and something in the kitty toward engine overhaul.

Fly Babies are about as simple as an airplane can be. They usually have small Continental engines. They're made of wood and covered with Dacron fabric. N500F didn't have an electrical system—no generator, no starter, no battery, no regulator, no solenoid, no radio, no transponder. We shared an open hangar with another homebuilt and paid $95 a month. Simple liability insurance cost $300 a year. State annual registration costs another $35. A good friend with an Airframe and Powerplant (A&P) license performed the annual inspection every year for free, with club members doing the actual repairs and maintenance.

In the eight years of the club's existence, our major maintenance expenses were:

1. A magneto coil: $85

2. A set of tires: $300

3. A replacement inner tube: $80

4. An oil cap: $20

5. Miscellaneous oil seals: $20
6. Repairs from wind damage: $150
7. A replacement windshield: $20
8. An exhaust pipe ($10)

Add that up, and you'll see the plane's average maintenance ran only about $100 or so, per year. Since the aircraft was operated in the experimental (amateur-built) category, we were able to save money by using parts from nonaircraft sources (the exhaust pipe came from an auto store) or make parts ourselves (like the windshield). The airplane ran quite happily on car gas, burning about 5 gallons an hour. Add 50 cents per hour for oil, and $5 per hour for the maintenance kitty, and you see where our low per-hour rate came from.

Mr. Bowers owned the airplane, and let our club operate it for free. But in 1994, he sold the plane. The members sadly waved good-bye, and disbanded the club.

Moonraker: Other decisions, higher costs

Two years later, I missed Fly Babies so much, I bought my own.

N45848 was 20 years newer than N500F, having first flown in the spring of 1982. Compared to N500F, it was luxuriously equipped—it had an electrical system, a radio, a transponder, a starter, navigation lights, and a cabin heater. It cost me $10,000, and had less than 100 hours total time with just 25 hours on its freshly-overhauled 85-horsepower Continental.

The direct operating costs are the same. I still fly on auto gas, but instead of paying $5 per hour into a kitty for eventual overhaul, I just figured to pay for it when the time came. My fixed costs, though—that's a different story.

It didn't start out that way. Originally, I kept the plane in the same open hangar, sharing with another plane. Eventually, though, my hangar partner moved out. I liked the extra room, and decided to keep the whole hangar for myself. So my monthly hangar rent jumped to about $190 per month (adjusted for inflation).

My new plane—dubbed "Moonraker" after the name for the highest sail in a square-rigged ship—turned out to have higher upkeep, too. On my first annual, one of the exhaust valves was bad. Getting it replaced cost about $400. In the nine years since, I've had to have the generator rebuilt, replaced the battery twice, replaced both the comm radio and the transponder, and replaced a faulty starter clutch ($350). Even when I write this, the plane is sitting in its hangar with the comm radio busted (again).

Annual maintenance costs have otherwise still averaged about $100. My A&P friend did the annuals for free for the first several years, but eventually retired. I now hire a mechanic for the annual, which adds another $250 to my yearly expenses.

And, in the interim, I got lazy. After doing some work on the plane in the open hangar with a driving rain blowing in and the roof leaking like a sieve, I signed onto the waiting list for some brand-new closed hangars at my airport. A year later, my Fly Baby moved into a large all-metal hangar with a locking door and access to a heated rest room (versus the sani-can at the old location). This came at a price, of course, almost $400 a month for the hangar.

Per-hour comparison

As you might imagine, the higher maintenance costs and more expensive hangar had a significant effect on the hourly rate, especially since I only fly about 35 hours per year. My rate in the club came to about $30 per hour. If there hadn't been any other members, the rate would have been about $60 per hour. With the new plane in the fancier hangar, and the higher maintenance costs of the more-complex Fly Baby, I pay about $175 an hour. That's almost twice what Cessna 172s rent for, at the same airport.

Am I nuts?

Well, yes: I'm nuts about airplanes, and Fly Babies in particular. While I have a larger-than-usual pack of other vices, none are costly. My income is sufficient to cover owning the airplane, and that's all I'm worried about. I didn't even *know* my per-hour rate was so high, until I computed it for this book.

The vast majority of my expenses are fixed costs—the hangar and insurance. My direct costs (fuel, and others) are trivial in comparison. I could switch back to the cheaper hangar, or find someone to share the one I'm in, and save almost $70 an hour.

But—for now, I won't. I feel that the radio and transponder in my open-cockpit aircraft are a lot less likely to disappear if I remain in the closed hangar with the locking door, and I like being able to disassemble my plane and leave it without worrying about discommoding a hangar-mate.

A summary of the two situations can be seen in Fig. 1-11, and further details on the Bowers Fly Baby can be found at *www.bowersflybaby.com*.

Of course, my situation isn't too typical. Let's take an in-depth look at what it costs to own a more-normal airplane.

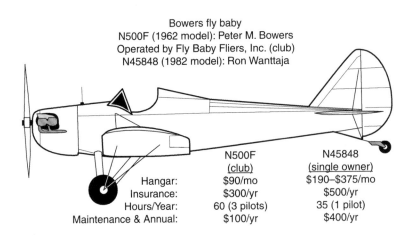

Bowers fly baby
N500F (1962 model): Peter M. Bowers
Operated by Fly Baby Fliers, Inc. (club)
N45848 (1982 model): Ron Wanttaja

	N500F (club)	N45848 (single owner)
Hangar:	$90/mo	$190–$375/mo
Insurance:	$300/yr	$500/yr
Hours/Year:	60 (3 pilots)	35 (1 pilot)
Maintenance & Annual:	$100/yr	$400/yr

Fig. 1-11. *A summary of Fly Baby costs for two different aircraft and two different ownership situations.*

2

The Costs of Ownership

Let's get down to dollars and cents. This chapter takes an in-depth, detailed view of the costs of aircraft ownership, and suggests ways of reducing them.

There are two types of costs incurred by the aircraft owner: fixed and direct. Fixed costs are those, which are connected with the *ownership*, and not the operation, of the aircraft. They're predictable, for the most part. Fixed costs are the most irritating aspect of aircraft ownership, as they continue whether the plane flies or not. It may snow 30 days out of the month, and on the thirty-first the driveway has to be shoveled, but you still have to pay the hangar, insurance, and other fixed costs. For the average owner flying a 100 hours or so a year, fixed costs are the major expense of ownership!

Fixed costs add up from a number of sources:

1. Finance costs

2. Insurance

3. Hangar/tiedown

4. Maintenance

5. Miscellaneous (taxes, fees, and the like.)

Direct costs are those that change depending upon how much the aircraft is flown. This mainly consists of the fuel and oil expenses.

Maintenance costs are really kind of a hybrid between fixed and direct. Your plane will require an inspection every year, so that's sort of a fixed cost. Yet the amount it costs may vary. Similarly, the more you fly the plane, the more things will break down and need repair.

Let's take a closer look.

FINANCE COSTS

How are you going to pay for your dream machine? If you have to take out a loan to buy your plane (Fig. 2-1), the monthly payments will probably form the largest single cost of ownership. If at all possible, save up for it rather than borrowing. A loan will add to your expenses in more ways than just principal and interest.

Fig. 2-1. *Few owners are able to pay cash for newer aircraft like this Columbia 400, so monthly loan payments become part of the costs of ownership. (Courtesy Columbia Aircraft.)*

Aircraft loans work similar to those for cars—the financial institution requires that you carry enough insurance to pay off the loan if the vehicle is damaged. The result is a higher insurance premium.

This won't happen if you finance some other way, like a signature loan. But their interest rates are higher. A home equity loan is an attractive possibility, but your spouse may balk at the thought of the family getting evicted if you can't make your airplane payments. Anyway, you'll still have monthly checks to write. It's easy enough to calculate what your monthly payment will be—just enter "mortgage payment calculator" or "loan payment calculator" as an Internet search term, and choose from several calculators.

Don't think you get off the hook if you pay cash, though. If you talk to an accountant, he or she will tell you that paying cash for your airplane costs money, too. They call it "lost interest" or "opportunity income." If you spent $40,000 on the plane, that money could instead have been invested and earned interest. To them, this represents lost income; hence, it's the same as an expense.

Whether to consider lost interest as a real cost of ownership is left to the reader. The more money that is committed to the aircraft, the more serious a factor it is. It's not that major with my $10,000 homebuilt, but someone contemplating a $120,000 Mooney had best keep it in mind.

Reducing finance costs

The most obvious way to reduce the finance costs is to win the airplane in a contest, right? Even if we indulge that fantasy for a moment, let's not forget the *taxes* due after such a windfall (Fig. 2-2). Winning a $200,000 plane would result in about a $70,000 tax bill; if you're like me, you'd have to take out a loan to pay it. You might end up paying $1000 a month finance costs on your "free" airplane.

On the more realistic side, paying cash is the least expensive solution. Maybe you've got a nice little nest egg tucked away. Maybe a small windfall drops into your lap. A friend sold his house and moved into an apartment to free up the necessary

Fig. 2-2. *Margaret Puckette won an Archer in an AOPA contest, but had to pay nearly $30,000 in additional income taxes.*

cash. Or perhaps you just start saving. If you can afford to *own* a basic airplane, you can save up the amount to *buy* the airplane in just a few years.

Need convincing? Let's take my old Cessna 150 as an example. Back in the mid-80s, it cost around $2000 a year to own and fly it. Yet I only paid $6000 for the aircraft. If I hadn't had the cash to buy it outright, I could have saved up enough in just three years.

Sure, this is kind of an unusual case. And you may not want just a little two-seat puddle jumper. But by delaying your decision for a couple of years, you can save a tidy sum toward the purchase of your dream machine.

What you have to do is, start saving. Now.

HANGAR/TIEDOWN

Whether you win, pay cash, or finance your plane, you're going to need somewhere to keep it. Other than loan payments (if any), hangar rent will hog the lion's share of your fixed costs. Tiedowns are cheaper, but hangars provide better protection.

A couple of different types of hangars are available. The most common pattern is called the T-hangar. The aircraft are aligned in long rows, with the tails of another row tucked between the planes.

A hangar might be nothing more than a long roof with no walls. These protect the planes from sunlight and precipitation, but provide no protection from the wind. Sometimes, the individual sections are walled off. Add a door, and you end up with a closed hangar that protects the airplane from the elements as well as van-dalism. Such protection costs more. It may pay off, though. Our club Fly Baby was kept in an open hangar, and I arrived at the airport one day to find all the propeller nuts missing. Somebody had tried to steal the prop.

There are permutations to the basic hangar pattern. Sometimes, the open T-hangar will have an exterior wall but no internal partitions. You're protected against wind and vandalism (the threat from your fellow owners is small). Or sometimes only partial external walls are erected. These provide some windbreak and provide places to set a storage cabinet.

Another hangar type crams a bunch of planes under a big free-standing roof. We'll call these "communal" hangars. These are often heated, which is a heck of a lot better on your aircraft. You'll pay extra, of course. Plus, your plane usually has to be moved to let some other plane out of the hangar. This leads to *hangar rash*—minor cosmetic damage occurring when wings bump into your plane. Plus, in some cases, you'll have to pay a fee to have your plane moved out of and into the hangar.

Whatever type of hangar you find, it's going to cost. The smallest closed hangars at my airport rent for $230 a month—a steal compared to most major metropolitan airports. Hangars 8 miles away at a controlled field run $500 on up. In a year, that's $6000. If you fly 100 hours a year, the hangar rent alone amounts to $60 an hour!

If your plane can withstand sitting outside, tiedowns are cheaper. The quality varies, just like hangars. The lowest-cost is a spot in the grass (Fig. 2-3). You'll probably be expected to install your own tiedown-rope anchors. On the opposite end of the scale will be a marked stall on a concrete apron, with solid attach points for your tiedown ropes. You'll find combinations, as well as an asphalt square for your airplane to sit, with grass taxiways, for instance. This isolates your plane from the moisture, but soggy ground can still affect your operations.

Rent varies, depending upon the desirability of the location. Like hangars, tiedowns cost more at close-in major airfields with good security. A tiedown may run $100 a month at these fields, but a grass strip 20 miles away may have spots for $35. This is illustrated in Fig. 2-4, which shows the location of various airports around Seattle, Washington, and their rental rates for both tiedown spots and hangars.

Whether an open tiedown works for you depends upon the severity of the local weather and the exterior surface of your airplane. With their aluminum skins, garden-variety Cessnas and Cherokees are good candidates for outside tiedowns. Fabric-covered planes are less suited. While modern synthetic covering materials last far longer than traditional fabric, they'll last a lot longer under a roof. Older tube-and-fabric aircraft often have spars or other components made of wood. You'd just as soon keep those out of the rain. Grit your teeth and get a hangar.

The cost of renting a hangar may be moot if none are available. The waiting list in my area is two years long. At one local airfield, closed hangars almost never become available. If you're seriously considering buying an airplane and you think you'll want a hangar, get your name in *now*!

Fig. 2-3. *The cheapest tiedown location is on grass, especially on out-of-the-way airports.*

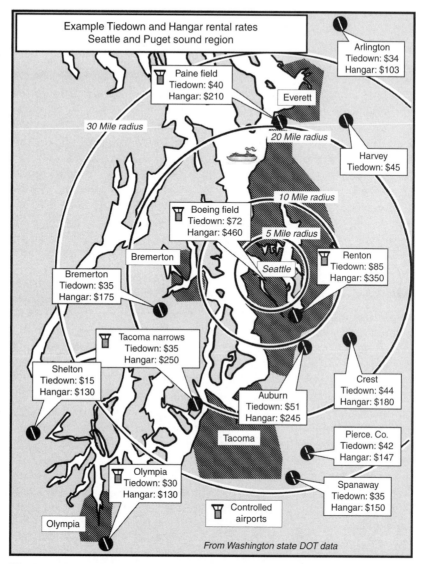

Fig. 2-4. *Hangar and tiedown costs will decrease in proportion to the airport's distance from major metropolitan areas.*

Fortunately, it's not difficult to get a handle on how much you'd have to pay for a tiedown space or a hangar—just call the local airports.

Saving money on hangar/tiedown

The best way to save money on tiedown or hangar rent is to shop around. As mentioned above, you can usually find the cheapest prices at the outlying, small fields. But how inconvenient is that to you? When are you going to fly? If you live right across the street from a major airport, keeping your plane 20 miles away just to save money might be false economy. If you can afford it, bite the bullet and keep the plane as close as possible to home.

I used to live about 15 minutes from the airport. About eight years ago, I moved to a new house just 5 minutes from the field. The advantage is striking. I can go to the hangar and "putter" whenever I have a few free moments. If I'm needed at home, my wife just calls me at my cell phone.

But if money is an issue, do some checking on smaller, private fields in your area. Back when I was learning to fly, our Civil Air Patrol (CAP) unit kept its Citabria in a barn on a cropduster's private strip. The door opening was actually narrower than the aircraft's wingspan. It took a little tricky maneuvering, but the rent was very low.

One excellent option for reducing hangar rental is to share the space with another airplane. Hangars are designed for a particular size of aircraft; often two smaller planes can be placed in an area designed for a mid-size twin. You have the problem with hangar rash mentioned earlier, but since it's just you and one other owner, you'll probably be a bit more careful. We'll talk about this more in Chap. 10.

TAKING A BIGGER PLUNGE

If you're willing to spend a good bit of change, you have two additional options: A "condo" hangar, or a home on a residential airpark.

Condominium hangars are being built at many airports. Instead of paying rent, aircraft owners actually purchase and own their hangars. In a few cases, the owners can build whatever type of hangar suits them. But in most cases, the condo hangars are just ordinary closed T-hangars identical to the usual rental units. While the monthly mortgage payment may approximately be what the owners might otherwise pay in rent, the hangar owner is at least building equity.

There are potential problems, of course. Owning your hangar gives you less flexibility if finances turn tight. Needing to sell both the plane and a condo hangar will add to your worries. If you don't own the hangar, you can tie the plane outdoors for a significant drop in the fixed costs. In most cases you don't have title to the land the hangar is built upon. A long-term lease is the usual practice, but some leases do stipulate that the entire structure be turned over to the lessor at the end of the lease.

The biggest plunge

The biggest step: Buy a house and hangar on an airport community, or "airpark." This is absolutely the best step for real airplane nuts. I've got several friends who own homes in local airparks, and they love them.

Drawbacks? First, if you still have to commute to work every day, you're probably faced with a pretty long drive. Airparks are usually well out in the country. This generally means less access to normal urban amenities like grocery stores, shopping malls, and the like. Realtors stress "Location, location, location" when shopping for a home, and while it's fun to have the plane just outside the back door, one still has a large involvement in the nonflying world.

Second, you're generally getting less "home" for the money. One couple I know has a home on a local airpark. They've got a 3000 square-foot hangar, but the attached house is only 1400 square feet, scarcely bigger than a large apartment. No kids, so they can get away with it, but the money they spend on their airpark home could have bought a nicer house a lot closer to town.

INSURANCE

Aircraft insurance has a lot of similarities to car insurance, but there are some critical differences. Just as with your car, the premium you'll pay to insure your aircraft will depend on a number of factors.

Take experience, for example. If you just got your driver's license, your auto coverage costs more. Similarly, if you don't have many flying hours, your aircraft premiums will be higher.

How about the aircraft itself? A Volkswagen Bug costs less to insure than a Porsche; and an Ercoupe's insurance bite will be less than that of a Cirrus.

Where's it based? If you live in a rough neighborhood, your car premiums are higher. And if you park your Bonanza at an open tiedown on a near-deserted strip, you'll pay more than if it's locked securely in a hangar on a security-patrolled airport.

Finally, how's your record? If you've had a lot of fender-benders and speeding tickets, you'll have trouble even getting automotive coverage. And if you do, it may be of the "preferred risk" variety; that is, expensive. Still, most drivers can't escape a little crumpled metal over the years, and a ticket for 37 mph in a 25 mph zone doesn't make you a scofflaw. "Honest, officer, my foot slipped...."

But that doesn't hold true in aviation. An FAA violation is a pretty big deal. Accidents are even rarer. The vast majority of pilots have neither bent aluminum nor voided licenses in their pasts. Those who have, stand out like sore thumbs and are subject to higher premiums.

Finally, your rates will depend upon how much insurance you carry. This is just like car insurance; basic liability is the cheapest. "Comprehensive" covers damage while the car is parked, and "collision" insures against accident damage.

Aircraft liability policies are similar to those of cars. Physical damage is handled with what is called "hull" insurance. Liability policies might cost $250 to $500 per year. Hull coverage runs between 1 and 10 percent of the value of the aircraft per annum. The higher figure is mostly reserved for high-risk aircraft like floatplanes; typical landplane premiums are 5 percent or lower.

Saving money on insurance

The type of aircraft you eventually select will have a strong effect on the insurance rates. A four-seat retractable taildragger will have higher premiums than a two-seat fixed-gear trigeared trainer. Of course, the two-seater might not be the right airplane for you. Still, some fixed-gear planes, like the Cirrus SR-20 or the Grumman Tiger (Fig. 2-5), can match the cruise speed of many light retractables. So don't reject the straight-leg planes out-of-hand.

Your premiums will also be based on a number of factors; some of which you can't affect. Fundamentally, premiums are set by how many flight hours you have and what ratings you hold. A 50-hour student pilot will pay more than a 1000-hour private pilot with an instrument rating. If you're ready to buy a plane now, there isn't much you can do.

Some companies reduce your premiums based on recurring training—the FAA's "Wings" program, for instance. Some, also, take pilot currency into account. If you fly 200 hours per year, an underwriter should consider you a better risk than someone who flies an hour a month. The plane is subject to risk more often, but a pilot with plenty of flight hours is more likely to avoid problems.

The fundamental solution for lower rates is to shop around. Don't be shy about calling a number of companies. Pick the general type of aircraft you think you want to buy, and ask for a quote based on your current pilot qualifications.

Hull coverage has a bit more leeway for minimizing costs. Premiums will be lower if the aircraft is kept in a locked hangar than tied down outside. You can also save dough by selecting a Hull policy that doesn't cover the aircraft while it's flying. This "Not in Flight" coverage is designed to give the owner some protection from hazards, which have nothing to do with aviation itself. An example would be avionics theft, hail damage, or a hangar fire (Fig. 2-6). If you're willing to bet on your piloting skills, a "Not in Flight" policy can save a significant portion of your insurance bill.

However, if you take out an aircraft loan, the lender will require your plane carry full Hull coverage.

Complete insurance details are described in Chap. 8.

Fig. 2-5. *This Grumman Tiger is almost as fast as some light retractable-gear aircraft, yet its insurance premiums are less.*

Fig. 2-6. *A leaky fuel system ignited this fire before the engine was even started. Damages would have been covered under a policy with "Not in Flight" hull coverage.*

MAINTENANCE COSTS

I wish I could tell you what the annual maintenance expense of your plane will be. But my crystal ball has been foggy lately.

Some costs are somewhat predictable. The annual inspection can run from a couple of hundred dollars to several thousands, depending upon the complexity of the aircraft. This is, of course, also affected by how many faults the inspection reveals.

For those that prefer a little predictability to their expenses, "flat-rate" annuals are available. Unless your plane requires extensive, unexpected work, the mechanic performs the inspection and minor repairs for a price agreed-on in advance. The price includes the labor for the inspection itself, and, usually, a bit of extra labor to fix the garden-variety minor problems. But if something big turns up, you'll pay for it. Parts aren't usually included in the price, either.

Flat rate annuals have both supporters and detractors. Some claim the operator will tend to skimp; to minimize the labor spent on inspection and repair. Others feel it keeps mechanics from repairing numerous small defects that don't affect airworthiness, but can be used to run their profits up. It's something you'll have to make your own decision on.

The annual isn't the only maintenance expense. Oil changes are required at regular intervals (generally 25–50 hours). Emergency Locator Transmitter (ELT) batteries have expiration dates and must be periodically replaced. Ditto for certain types of hoses. Transponders must be tested and inspected every two years, as must the altimeters and static system for aircraft flown under Instrument Flight Rules (IFR).

As far as figuring out the cost of maintenance for your prospective purchase, the complexity of the aircraft must be taken into account. A Bonanza, for instance, is going to cost more to inspect and maintain than a Cherokee 140. Every year, for instance, the inspector will have to place the Beech on jacks and test the retraction system (Fig. 2-7).

Fig. 2-7. *The FAA requires that the gear mechanisms on retractable-gear aircraft be tested at every annual. This increases the cost of the inspection.*

But maintenance-wise, annuals and routine costs aren't the big worry. Occasionally things go bad. Very bad. Years ago, a friend had to replace two engine cylinders in his Tri-Pacer. The total cost was about 15 percent of the value of the plane.

It isn't just parts wearing out, either. Manufacturer service bulletins and FAA Airworthiness Directives (ADs) can add to unanticipated maintenance expenses. Aircraft builders track problems their planes have. If a certain problem keeps recurring in a particular model, they sometimes will develop an improved part. The availability of the part and the underlying problem will be announced as part of a service bulletin. Compliance with these bulletins is optional. However, if the problem is especially safety-related, the FAA may issue an AD. Compliance with an AD is mandatory.

The best way to handle the unexpected maintenance expense is by establishing a maintenance reserve by setting aside a fixed amount every flight hour. The basic purpose of this fund is to pay for the eventual overhaul of the engine. It also comes in handy for other expensive repairs.

Determining the rate to "charge" yourself for the maintenance reserve is relatively easy, it's based upon the Time Between Overhauls (TBO) of the engine and the cost of overhaul. Typical engine TBOs range from 1600 to 2200 hours. To determine the hourly rate for the maintenance reserve, take the overhaul cost and divide it by the hours until overhaul (at the time you bought the plane). It's as simple as that. The Lycoming O-360 shown in Fig. 2-8 has a 2000-hour TBO. Count on an overhaul cost of $10,000 (rather low, really) and build your maintenance reserve at the rate of $5 a flight hour. If you have a Lycoming with 1000 flight hours, the rate jumps to $10 an hour, since the overhaul charge must be "earned" over only 1000 flight hours.

The maintenance reserve should be deposited into a separate account to keep it isolated from other needs. Of course, it is available for worthier causes: if the spouse simply must have new bedroom furniture, so be it. Or, you may decide to ignore the maintenance reserve. Most owners don't keep their aircraft that long; you'll probably sell the plane long before overhaul time. Being more grasshopper than ant, my intent has always been to dig up the money from somewhere when the time comes.

Fig. 2-8. *The Lycoming O-360 engine is used on a variety of aircraft. (Courtesy Textron Lycoming.)*

Keep in mind that TBO is a rather nebulous term. The ins and outs of engines are discussed later.

With ordinary luck, there shouldn't be many unexpected maintenance hits on your budget. When you read the various case studies, you'll note that the annual unexpected maintenance costs range from $100 to $6000.

Saving money on maintenance

If you've got money to burn, go ahead and have an A&P mechanic do all the maintenance on your airplane. But if your tender's too tight for tinder, do as much of the work as you can.

The Federal Aviation Regulations (FAR) state: "The holder of a pilot certificate... may perform preventative maintenance on any aircraft owned or operated by that pilot which is not used under Part 121, 127, 129, or 135." That's from FAR 43.3. The term "preventative maintenance" is defined in Appendix A of FAR 43, and summarized in Fig. 2-9.

If you glance at the list, there's a lot of stuff you can legally do! Change the oil. Replace tires and tubes. Grease wheel bearings. Patch fabric and metal. Replace bulbs, and clean and gap spark plugs.

How much can you save? Let's just take "change oil" for an example. A small Continental engine's oil must be drained and replaced every 25 flight hours. It'll probably cost you about $70 to have that done by an A&P. Yet the oil itself is $3 or so a quart. For a fun afternoon of messing with your plane and getting greasy, you've saved about $50. If you fly 100 hours a year, that's over $200 you've saved on oil changes alone. And that's just one aspect of owner maintenance.

Chapter 12 gives you the background necessary for performing common preventative maintenance. Of course, there are a lot of other items that must be left to licensed mechanics. The annual inspection is the big one. But consider the "owner-assisted" annual. The inspection involves a lot of "grunt" work that doesn't require

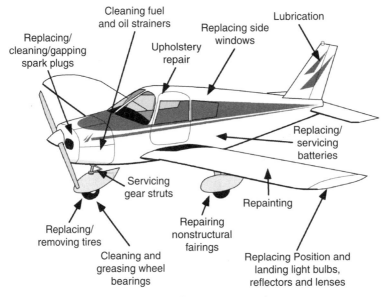

Fig. 2-9. *FAA-Approved owner maintenance operations.*

a licensed mechanic to perform. Many facilities reduce the price if the owner agrees to perform these simple tasks. If you participate, you'll do jobs like opening and closing inspection panels, removing the seats and carpeting, washing the exterior and vacuuming the inside, lubricating hinges, changing light bulbs, fetching parts, holding propellers and flashlights, and being a general maid-of-all-work.

Owner-assisted annuals are wonderful opportunities to see how your airplane works and to learn the proper procedures for aircraft maintenance. They're cheaper, too! However, not all facilities offer them. Plus, the FBOs generally only operate Monday through Friday, so you'll have to take time off work to help.

Finally, it doesn't hurt to shop around when selecting A&Ps to work on your plane. Often, mechanics in the small, outlying airports have lower rates and are more willing to allow the owner to participate.

DIRECT COSTS

Direct costs are easy to compute. Take the engine's per-hour fuel burn and multiply by the cost of fuel. Add a buck or so for oil, and there are the direct costs. But there is one major variable in the direct cost equation: The cost of gas. The solution is simple: Buy an airplane with an engine capable of running on 80 octane avgas. Of course, 80 octane is rare these days, but autogas isn't. Most 80-octane airplanes can be approved to run on ordinary car gasoline. As I write this, unleaded car gas is $2.20 a gallon, while 100LL is $3.50. Plus, I can apply for a 10-cent per gallon refund on the state road taxes included at the pump.

There are drawbacks. The logistics can be frightful, depending upon your aircraft. It's not as hard to haul gas for a Luscombe than it is to keep a Cessna 182's tanks brimming with autogas. While some airports sell auto fuel on-field just like avgas, you'll probably have to haul gas from the local service station. Carrying it and filling your tanks are fraught with obvious hazards.

To legally use car gas, the aircraft must have received a Supplemental Type Certificate (STC). The Experimental Aircraft Association (EAA) and Peterson Aviation have performed tests to the FAA's satisfaction on dozens of aircraft types. If your plane doesn't have the STC, you must purchase the STC from either EAA or Peterson. For most airplanes, the only modifications required are a logbook entry and the replacement of the "Fuel Grade" sticker by the gas cap.

What will happen if you use car gas without the STC? If the FAA catches you, it's a violation. Unfortunately, the way they usually catch violators is in a postcrash fuel analysis. Probably your biggest worry is your insurance company. If you have an accident and you were using autogas *without* the STC, the insurance company may be able to deny your claim.

The other major direct cost is oil. Engines typically burn a quart every 5 or 10 hours. The oil should be changed every 25 hours for engines without oil filters, or at 50-hour intervals if an oil filter is installed. Aviation oil costs two bucks or so a quart, hence the oil costs come to about a dollar an hour.

ADDING IT UP

Table 2-1 shows typical costs based on various factors. Five cases are presented:

- Case 1 presumes a light two-seat airplane like a Champ or Cessna 150. Only liability insurance is carried, and the plane is run on auto fuel. It's assumed you buy the plane with cash and don't have to make payments on a loan. The plane has an outside tiedown.

- Case 2 presumes a typical four seater like a Cessna 172 or Piper Warrior, purchased with cash as in Case 1, and also running on auto fuel. It's kept in an open T-Hangar.

- Case 3 is the same airplane as Case 2, only it includes monthly payments on a $25,000 loan.

- Case 4 is the same airplane as Case 3, only it is run on aviation fuel instead of car gas.

- Case 5 is a light retractable like a Mooney or Piper Arrow, in a closed hangar moderately close to an urban center.

Cases 2, 3, and 4, presenting the same type of plane in slightly different situations, provide good insight. Notice that the addition of loan payments (Case 2 to Case 3) essentially *doubles* the per-hour cost of the aircraft! The use of auto fuel vs. 100LL saves about $10 per hour. This is a significant percentage of the total cost for the cases where a loan isn't necessary, but once one is making those ~$400 payments per month, the autogas savings rather pale in comparison.

As discussed in the last chapter, an annual utilization rate of ~125 hours per year is about the break-even point between owning and renting a Cessna 172-class airplane. As mentioned, though, if you want to own an airplane, the primary point

Table 2-1. Sample costs of ownership

Examples of Ownership Cases

	Case 1	Case 2	Case 3	Case 4	Case 5	Cost Assumption
Storage						
Tiedown	X					$40/mo
Open T-hangar		X	X	X		$150/mo
Closed hangar					X	$300/mo
Loan amount	$0	$0	$25,000	$25,000	$50,000	
Loan term	—	—	7	7	10	7% Interest
Fuel						
Autofuel	X	X	X			$2.30/gal
Avgas				X	X	$3.50/gal
GPH	5	8	8	8	12	
Seats	2	4	4	4	4	
Gear	Fixed	Fixed	Fixed	Fixed	Retract	

Monthly Expenses

	Case 1	Case 2	Case 3	Case 4	Case 5
Storage	$40	$150	$150	$150	$300
Loan	$0	$0	$4560	$4560	$6960
Maintenance	$50	$75	$75	$75	$100

Yearly Expenses

	Case 1	Case 2	Case 3	Case 4	Case 5
Liability insurance	$500	$700	$700	$700	$700
Hull insurance	—	$500	$500	$500	$1000
Annual inspection	$500	$1000	$1000	$1000	$2000

Total Annual Cost Summary

		Case 1	Case 2	Case 3	Case 4	Case 5
25 Hours/year	Fuel cost	$288	$460	$460	$700	$1,050
	Total cost	$2368	$5360	$9,920	$10,160	$16,510
	Per hour	$95	$214	$397	$406	$660
50 Hours/year	Fuel cost	$575	$920	$920	$1,400	$2,100
	Total cost	$2655	$5820	$10,380	$10,860	$17,560
	Per hour	$53	$116	$208	$217	$351

100 Hours/year	Fuel cost	$1150	$1840	$1,840	$2,800	$4,200
	Total cost	$3230	$6740	$11,300	$12,260	$19,660
	Per hour	$32	$67	$113	$123	$197
200 Hours/year	Fuel cost	$2300	$3680	$3,680	$5,600	$8,400
	Total cost	$4380	$8580	$13,140	$15,060	$23,860
	Per hour	$22	$43	$66	$75	$119

of interest in Table 2-1 is the total annual cost. If you can afford the total cost, it doesn't really matter what the per-hour rate is.

Do the numbers shock you? Are they completely out of reach? If so, stay tuned: There's a way to cut those numbers by 50 percent or more.

PARTNERSHIPS: SHARING THE FUN

Assume you've worked out that your dream plane will cost about $1000 a month to own. Too much? Well, if you bought *half* the airplane, your expenses would drop drastically, wouldn't they? All you have to do is find someone to buy the other half of the airplane.

It makes a lot of sense. Most of the expense of ownership comes from the fixed costs, not the direct ones. Let's say you fly 100 hours a year, and your fuel cost is $2000. Assume that the fixed costs for the plane run another $5000. If you owned the plane yourself, the plane would be costing you $70 per hour.

With a single partner, your part of the annual fixed costs drop to $2500. You'd still be paying the same direct costs, but your plane just got $25 per hour cheaper to run! With a third partner, that drives your rate down to about $36 an hour—about half the original amount.

On the downside, the airplane's not exclusively yours. That can dull the cachet for some. But it's still better than renting. Do you need the plane for a whole week? Just call up your partners and clear it with them. Have to leave it at another field due to bad weather? Ditto. And if an unexpected maintenance bite hits, the cost gets spread out across many shoulders instead of just yours.

Partnerships versus clubs

The terms *partnership* and *club* are sometimes used interchangeably. The difference is subtle, yet important. In a partnership, the participants actually own a portion of a single aircraft. They pay a monthly charge based upon the actual cost of upkeep for the aircraft (hangar, insurance, reserve for overhaul, and the like.) If an unexpected expense occurs, the members chip in equally to handle the bill. All decisions are made cooperatively by the owners, which generally number four or less. The only way out of the arrangement is to sell their portion of the plane, usually under some restrictions place by the partnership agreement.

In a flying club, the organization itself owns the aircraft. Membership is achieved by paying an initiation fee, which may or may not be refundable upon leaving the club. Decisions are made by a duly-elected board. Members pay a fixed monthly fee. The ratio of aircraft to participants is usually higher (one local club aims for 10 members per plane), and the club may own multiple aircraft.

Either way has both advantages and disadvantages. With a partnership, you can point at the plane and say, "That's mine." But if a major maintenance bill arrives, that's *also* yours—at least part of it.

Individual expenses will be far less in a club. Yet the pride of ownership isn't quite there. If you're leaving the FBO to escape scheduling problems, a club may not offer much relief.

Much of the following information applies to both arrangements. McGraw-Hill's *Aircraft Partnership*, by Geza Szurovy, is an excellent in-depth guide for those contemplating joint ownership. We'll cover partnerships, mostly, in the following sections, although many of the details apply to flying clubs as well.

Member friction

Obviously, finding the right person(s) to buy an airplane is paramount to the success of the endeavor. There are, potentially, a number of areas of friction:

1. *Fiscal irresponsibility.* There are bills that must be paid; paying them depends upon the participation of all parties concerned. If your partner gets behind on his portion of the expenses, you'll end up shouldering the burden. If the partner handles the financial side and neglects to make aircraft payments, that's *your* airplane that gets possessed.

   It's more than just dealing with someone who won't pay his or her portion of the regular bills. You may feel it's time for the partnership to dig down and get the engine overhauled. Your partner may feel quite comfortable with the current state of the engine. Or, your partner wants to get his instrument ticket, and feels the panel should be upgraded. Yet all you want is a simple VFR fun machine.

2. *Workload.* Owning and maintaining airplanes is a lot of work. A partnership setup can reduce its impact on you, that is, assuming the others provide their fair share. Like the monetary aspect, the load may end up in your court.

3. *Pilot irresponsibility.* You're driving along a country road, and you see the 172 that you own with three other partners. Suddenly, the Cessna's nose dips, then pulls up into a broad loop.

   What are you going to do about it?

   Or perhaps you show up at the airport to go flying, just as your partner taxis in. You take off, but upon landing find the FAA waiting. Seems a plane buzzed a local beach, and someone wrote down your N-number. You show the FAA the notebook where each member is supposed to record his or her flight time, only to discover that the earlier pilot hadn't recorded his time. You're the only one (officially) who'd flown that day.

What are you going to do next?

4. *Incompatibility.* I can tell when another member has flown our club homebuilt—the tiedowns are attached slightly differently. I think my way is better, but don't make an issue of it. What if we disagreed on something a little more vital, such as engine operation? How could we resolve it?

The partnership agreement

One of the ways to avoid friction is to spell out the responsibilities and duties of each partner in a formal partnership agreement. Joint ownership of something as complex and valuable as an airplane should not be taken lightly. The partnership agreement should indicate the obligations and privileges of each member, and the actions to be taken during certain circumstances.

Some of the aspects that should be mentioned:

- How ownership expenses are assessed from the partners
- Who handles the finances, and at what level of outlay must he or she first consult the other partners
- Restrictions on use (commercial, training, and so on.)
- Responsibility for paying the insurance deductible in the event of an accident
- Process by which one partner can buy out the other(s)
- How the airplane is scheduled
- Arbitration procedures
- Insurance to be carried

Any number of other aspects should be addressed. I recommend that you don't try to assemble an agreement from scratch. Assemble several sample agreements and come up with one directly tailored to your situation. Szurovy's *Airplane Partnership* contains one such sample. In addition, Aircraft Owners and Pilots Association (AOPA) membership services can provide a packet on coownership, which includes another sample agreement.

This book contains two case studies which involve aircraft partnerships: The Stearman in Chap. 6, and the Cherokee Six in Chap. 8.

LEASEBACKS

Ever wonder *where* those planes at the FBO come from?

The FBO itself may own some. But others are probably owned by private individuals. They buy them, maintain them, and let the FBO rent them. In return, the owner gets a portion of the rental fee and preferential treatment for scheduling the aircraft for their own flying.

Leaseback arrangements are designed to benefit all parties. The FBO gets the use of an aircraft without the capital outlay of buying it. Pilots can fly an aircraft without the necessity of buying one. And the owner's expenses are alleviated by the rental fees paid by the pilots.

Alleviated, not necessarily *covered*. Significant tax breaks used to be available for owners who put planes on leaseback. These could convert a money-losing arrangement into one that at least breaks even. Unfortunately, most of those benefits were lost during the "tax reforms" of the early 80s.

Some advantage, taxwise, is still retained. When carefully set up, a leaseback will allow you to deduct the expenses involved in owning your aircraft.

Leasebacks can work out for an aircraft owner. Unfortunately, there are tons of variables out of your control. You can specify the minimum level of pilot qualification for renting your plane, but there's no way control how gently (or otherwise) they treat it. Does the leaseback agreement require you to have all inspections and maintenance done in the FBO's high-priced shop? How honest is the FBO? How *stable* is the FBO? What if you get committed to high monthly loan payments and the FBO goes bankrupt?

As a more solid companion to the what-ifs: You'll still be responsible for all the expenses of the aircraft. Your plane will require 100-hour inspections, which are practically identical to the annual. The more popular your plane is on the rental line, the more often you'll be paying for 100-hour inspections. Insurance premiums will be five (or more) times higher. You'll probably get preferential treatment when scheduling your own plane, but in many ways, it'll be just like renting.

Do your research. Talk to other owners who leaseback to the same FBO. Pick a popular aircraft, one that'll get a lot of rental time. Leasebacks aren't something to leap into. Much, but not all, of the feedback I got on leasebacks was negative. Yet some owners seem to make it work. Here's one who did:

CASE STUDY: A DIAMOND ON LEASEBACK

Eric Warren wanted an airplane—but not just *any* plane.

"Cessnas bore me," says the Houston-area pilot. "Pipers were just too small." He was still a student pilot, so planes like Mooneys and Beeches were past his skill level. The Cirrus aircraft were out of his price range.

He settled on the Diamond DA-40 Diamond Star, powered by the Lycoming IO-360 engine. "It was the only plane that was certified, fun to fly, a trainer, and still an all-around excellent plane." He decided on a leaseback arrangement as the only way he could justify the airplane for business; many of the aircraft expenses would then become tax deductions.

Finding the airplane was a stroke of luck. The local Diamond dealer had a 2001 Diamond Star as a demonstrator, but had lost their distributorship and were looking to sell the plane (Fig. 2-10). List price was about $190,000, but he dickered the owner down to $160,000. While Diamond aircraft had selling at discounted prices at the time, Warren still got what he calls "a killer deal." "$170,000 would have been more likely for that plane, if the owner hadn't lost his dealership."

Warren had been doing well in his job in the technology sales field, and amassed a good down payment from his commission checks. He financed the rest of the plane, ending up with a monthly payment of around $1325.

Fig. 2-10. *Eric Warren bought his Diamond Star DA-40 when the FBO losts its distributorship, then placed the plane on leaseback with the same FBO.*

Ownership experience

Warren negotiated the leaseback arrangement with the same FBO that sold him the plane. The FBO managed the rentals, scheduled the plane, and performed all the maintenance. The FBO collects the rental, keeping a management fee.

For the first year of the leaseback, Warren paid the FBO a nominal rental fee whenever he scheduled his own aircraft, but he didn't for the second year. "Some FBOs will talk owners into this for tax reasons, but it is not at all necessary. It's up to the owner and their accountant/lawyer." He did find that managing the leaseback arrangement took a surprising amount of his time.

During his two years of ownership of the leaseback airplane, Warren earned his private license, started on his instrument rating, and used the plane both for local sight-seeing and longer trips with his wife. The airplane was based at about a half-hour drive from home. It was tied down on the FBO's flight line, for which he paid $115 a month. On the leaseback arrangement, the plane flew about 250 hours per year.

Since the plane was operated commercially and flown by renter pilots, insurance was fairly steep: $11,000 per year. His first 100-hour inspection cost about $1,200, but he attributes part of that to "training the mechanic." Many A&Ps are not familiar with newer planes like the Diamond, and they may take a bit more time to do routine tasks. Later inspections dropped to about $600, with an additional $300–$400 in additional expenses. For instance, at one point, Warren did have to pay for some minor repairs to the Diamond's composite airframe. In another case, the FBO paid to repair damage to the aircraft door.

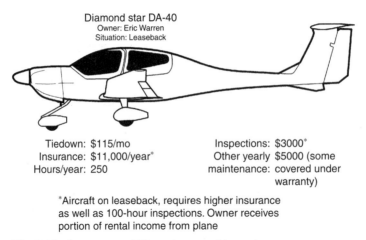

Diamond star DA-40
Owner: Eric Warren
Situation: Leaseback

Tiedown: $115/mo
Insurance: $11,000/year*
Hours/year: 250

Inspections: $3000*
Other yearly $5000 (some
maintenance: covered under
warranty)

*Aircraft on leaseback, requires higher insurance
as well as 100-hour inspections. Owner receives
portion of rental income from plane

Fig. 2-11. *A summary of Warren's ownership costs.*

His annual maintenance bill approached $5000, though some work was performed under warranty. "The exhaust systems were the weak point," says Warren, "But Diamond gave me *lots* of free parts. Even after the warranty expired, because it was a known issue." He did end up paying for about $800 in exhaust repairs, and had to replace the aircraft's starter at one point.

Warren's ownership experienced is summarized in Fig. 2-11.

Advice

Eric Warren bought the airplane to get his own license and gain experience toward later purchase of a high-performance aircraft. He put the Diamond Star on leaseback to enable ownership of a new, modern-technology aircraft.

He succeeded in both counts. He sold the airplane after two years, and now owns a 1995 Mooney Ovation.

On the business side, Warren feels he did a bit better than the break-even point. "Not counting taxes, I likely lost a good amount, perhaps $30,000. However, I had no other way at the time to justify the plane for business. Counting taxes, I'm really ahead by a few thousand." He sold the Diamond for just $13,000 less than he paid for it—less than 10 percent depreciation on a brand-new airplane over two years. That's pretty good!

His advice to those contemplating the same path: "When contemplating a leaseback and an FBO, you have to be realistic. Don't expect to make money, just expect to cover costs and have help managing the plane. Also, have money available, and realize that someone else my wreck it."

3

Picking the Plane

Of all the aspects of the aircraft-buying process, deciding *what* to buy is probably the most fun. Cessna? Piper? Cirrus? Beech? A traditional plane, or one of the new Light Sport Aircraft (LSA)? Now were talking about *airplanes*, instead of dry financial details.

But too often, an airplane is picked for the wrong reason. For example, a buyer thinks she needs a fast airplane, only to discover that the aeronautical hot rod can't handle the backcountry fields where the owner wants to pitch her tent. The result is an unhappy camper, or even a crumpled mass of aluminum when the owner tries to shoehorn the beast into the wrong field.

There's a lot of difference between a 1965 Cessna 150 and a 2006 Beechcraft—and I'm not just referring to the half-million dollar difference in cost. The difference is in the owner's *mission*: What they need the airplane for? Obviously, if you need a Bonanza, the 150 won't do the trick, but if all you need is the Cessna, you might have other uses for that $500,000 extra you'd have to shell out for the Beechcraft.

Sure, that's an extreme case. But the principle is the same—you don't want to spend any more money than you have to for a plane that meets your needs.

The first step is to determine your mission.

DETERMINING YOUR MISSION

There are as many missions as there are pilots. Some just like to fly for the pure joy of it—sometimes called "cutting holes in the sky." These folks could get by with a small, simple aircraft. The new LSAs are perfect examples. For all intents and purposes, these are just updated versions of Piper Cubs and Aeronca Champs (Fig. 3-1). And they didn't call the Champs "Airknockers" for nothing. For those who prefer metal skins and tricycle gear, the Cessna 150/152 can't really be beat for a cheap, simple aircraft.

Such basic planes aren't enough for some. They want room for radios and baggage, higher cruise speeds, and longer range. Some plan to operate from short grass fields, but others intend to use their planes for serious travel. By "serious," I don't

Fig. 3-1. *The Legend Cub was one of the first examples of the certified Special Light Sport Aircraft. A simplified certification process makes it easier to bring lower-cost ready-to-fly small aircraft to the market. (Courtesy Legend Aircraft.)*

mean to imply that the travel is for business purposes. Rather, the airplane is used as a mode of transportation rather than pure entertainment (Fig. 3-2). If it's more fun to take the Piper than the Pontiac, that's pure gravy.

Other needs exist. Perhaps you're interested in aerobatics, and a Super Decathlon (Fig. 3-3) seems like the way to go. Maybe you've got a yen for history that an antique or classic can scratch. Or maybe a floatplane or amphibian would be more your style. But is the mission really right for you? Will you outgrow the simple little "puddle

Fig. 3-2. *Those with serious travel needs should look at planes like this Saratoga II HP. (Courtesy Piper Aircraft Corporation.)*

Fig. 3-3. *The American Champion Super Decathlon is one of the few production aerobatic airplanes on the market. (Courtesy American Champion Aircraft.)*

jumper?" Are you really that interested in aerobatics? Do some serious thinking about how your needs might change.

Sometimes it's tough to determine a single mission. For example, you'd like a fun little knock-around aircraft, but still require fast cross-country travel. Keep in mind that good cross-country aircraft can be rented at practically any FBO. Buy the LSA for fun, and rent a Piper Arrow or equivalent for heavy-duty traveling.

To start, come up with a simple sentence or phrase expressing your basic mission, like, "Carry one passenger on long cross-countries between major airports." Or, "amateur competition aerobatics." "Operate from short private airstrip, with high-speed travel to nearby destinations." "Good trainer for kid's lessons."

Determining your mission may seem trivial, yet it is an important step in figuring out which plane fits your requirements for the least money.

The mission pitfall

You've determined your mission, and are ready to start scanning the performance charts. Before you do, take another look: Is your stated mission *really* what you need? Because if you buy *more* airplane than necessary, not only are you wasting money, but you risk reducing your flying enjoyment.

Say, for instance, your first cut at a mission statement was: "To fly two adults and two kids on long cross-countries." Now, that's pretty reasonable. You'll obviously need a four-seat airplane with a moderate amount of fuel capacity. Any number of planes would do—the Cessna 172 and Piper Warrior come to mind.

However, you know that a 172 with four people on board (even if two of them are kids) can get somewhat marginal on useful load. It's tempting to modify the mission statement: "To fly two adults and two kids and two hundred pounds of baggage long cross-countries." Oh-oh, that means you'll need a higher useful load. The plane has just moved into the Piper Archer/Cessna 182 (Fig. 3-4) category. Then you think, "Hmm, I might want to take the folks along." Up goes the mission statement to: "Fly *four* adults and two kids...." And you've just kicked the plane into the Cessna 207/Piper Saratoga category.

What's wrong with that? Nothing—*if* you really need the capacity. What is it costing? Money, of course. That Saratoga will cost a lot more than the Warrior you started with. It'll burn nearly twice the gas, and have a proportionately higher maintenance cost. Plus, if you really think you'll be doing most of your flying solo, the plane will be less pleasant to fly. Those load-haulers have a pretty broad Center of Gravity (CG) range. When flown solo, they'll be at the forward limits of their CGs. You'll have to horse back on the yoke to rotate, and you may not have enough elevator authority to keep from landing nosewheel-first. It might rob you of some of the fun of ownership.

You have to draw the line somewhere, otherwise, you could end up with a DC-3 instead of a Cessna 180. Look at your mission realistically, and try to look at those airplanes, which give just the right performance.

Ownership without guilt

"Mission?" some are wondering. "I just want to own a Hotrock 200!" In other words, you're not making a logical, rational decision based upon careful analysis of needs and features.

There's nothing wrong with that!

Few of us have to own an airplane. We've already established that ownership is a hobby; the airplane's not expected to earn its own way.

Fig. 3-4. *While the Cessna 172 and 182 have four seats, the larger 182 allows more baggage and fuel to be carried if all four seats are occupied. (Courtesy Cessna Aircraft Company.)*

However, there are things you expect to do with your Hotrock 200. Maybe your fantasy is flitting to that little mountain strip near your favorite fishing hole. Or, packing the wife and kids aboard and winging your way to Grandma's at 200 knots.

Before you spring the bucks, shouldn't you at least make sure the plane matches your needs?

One of the rites of manhood is the buying of the first car. Considerable effort goes toward finding exactly the right set of "wheels" to fit one's trans-pubescent self-image. When found, it is bought, whether Mom and Dad approve or not.

Reality soon intrudes. The engine leaks oil like a grounded tanker. The chassis is sprung, the shocks are shot, the electrical system shorts out. Have you ever seen a shiny older car at the side of the road, with a stricken teenager tenderly stroking the steering wheel, waiting for Dad and a tow rope? Too often, that first car is one's introduction to the need for rational decisions, instead of emotional ones.

Its bad enough with a $1000 car. Do you want to go through the same process with a $50,000 airplane? Shouldn't you make sure the plane is what you need, and not just what you want? But hey, it's your money and your decision. If your analysis finds Brand X better meets your needs than the Hotrock 2000, but you still prefer the Hotrock, go for it!

ESTABLISHING THE REQUIREMENTS

It's all well and good to make a highfaluting statement regarding what your airplane's going to be used for. Let's break the simple mission statement into hard requirements.

Performance

First, let's look at how much runway you have. If your mission calls for operation from "standard" airports only, there's no real takeoff/landing requirement. By standard airport, I mean those with 3000–4000 foot runways, fairly clear approaches, and so on.

Once you're off the ground, four interrelated parameters are cruise speed, range, fuel capacity, and fuel consumption. Since refueling operations severely impact total time enroute, set the endurance requirement (usable fuel divided by fuel consumption at cruise) first. Some planes have only 2 or 3 hours of onboard fuel; fine for local pleasure flights but serious cross-country travel requires more. Again, set the minimum value based on your mission. Be realistic, though, how long do you honestly want to fly without a potty break? A typical value might be 3 hours; add another 30 or 45 minutes for reserves and round it up to 4 hours.

Now state your desired cruise speed. Local sport flying, of course, doesn't need a cruise speed requirement. But don't get hung up on an absolute value. Say you want a 200 mph cruise. How much time is lost if you compromise on a plane 20 mph slower? Less than you'd think. Take a 600 mile trip. The first plane makes it in 3 hours, and the second one arrives only 20 minutes later. The slightly-slower plane may well be significantly cheaper, so keep flexible on your performance requirements.

In any case, take your endurance requirement and divide it by cruise speed to determine your aircraft's required range. If your requirements include operation from smaller strips, your selection process just becomes more difficult. Published performance figures don't really indicate, which airplanes are truly better at short-field work. For instance, both the American AA-1A (Fig. 3-5) and the Beech Bonanza list takeoff at distances of less than 1000 feet. Yet the Grumman has proven to be a poor short-field performer in comparison.

You'll have to dig a bit deeper for information if you'll be flying from shorter runways. Talk to owners and read some of the magazines and books, which review various aircraft.

Capacity

Among your mission statement should be a reference as to how much carrying capacity the plane should have. You might express it as "seats," or in a load amount, if you intend to carry something inanimate.

It's an unfortunate aviation fact of life that most planes can't legally, safely, and/or comfortably haul full fuel plus as many people as they have seats for. In many circles, a 172 is considered a three-seat, rather than four-seat airplane.

If you've got a definitive need for a plane that can carry four adults, shift your attention to the six-seaters. A four-seater might do OK with two adults and two kids, but baggage will complicate the issue... as will the fact that those kids are going to get heavier as they grow up.

If you are going to need to carry cargos, ensure the passenger seats on the planes you look at easily remove. The planes should also have good-sized doors for loading bulky objects, and anchor points on the floor to secure the load. Even if the plane has the right number of seats, consider how much room the plane gives you.

Fig. 3-5. *While the specified takeoff and landing distances for planes like this American AA-1A imply it can operate from short fields, the reality is somewhat different.*

A Cessna 152 is not the platform for two largish people, even if the plane stays within the gross weight limit. A Cessna 172 has not only an increased useful load, but a 6-inch wider cabin.

It's always nice to have some margin. If you're at a toss-up between a two- or four-seater, the larger aircraft is usually the best way to go. It'll give you the most flexibility and is quicker to sell when you put it on the market. Not much more expensive to operate, either. However, two-seaters are invariably cheaper to buy.

Preferences

Performance and capacity aren't the only issues. Your mission may require that the aircraft have certain features. Not to mention catering to your own biases.

What's your preferred configuration? High wing? Low wing? Side-by-side seating? Tandem? There are advantages and disadvantages to each.

Generally, a low-winger gives better flight visibility for maneuvering in crowded airspace. A high wing aircraft is better for sightseeing, and are generally easier to get into and out of.

How many doors? Some planes have only a single door for the cabin occupants. It's a lot easier with two, but doors can sometimes turn into maintenance problems. Some planes do without doors entirely (Fig. 3-6). These enhance the sporty feel of the airplane and make entry and exit a lot easier, but can be a problem if they're opened when the wind is gusty or it's raining.

Need IFR capability? You can turn just about any airplane into an IFR platform with enough money and a willingness to forego common sense. I've seen pictures of full-IFR Super Cubs and Waco biplanes. Yet there's more to it than mere panel space. Everyone agrees that the superior stability of a Piper Warrior makes it a better platform for instrument work than a Luscombe, for example.

Fig. 3-6. *Tip-up canopies on planes like this Diamond DA-40 make entry and exit a lot easier. (Courtesy Diamond Aircraft.)*

There are other panel questions as well. Flying in/around congested airspace? Multiple Comm radios will be handy. Mission call for long cross-countries? Put a moving-map Global Postioning System (GPS) on the wish-list.

If you do decide that a two-seater will fit your mission, be aware that the seat configuration plays a role in performance and pilot/passenger satisfaction. A tandem-seating aircraft (which has one seat in front and the second in back, like a Cub or Champ) gives the pilot good visibility to either side and places him or her on the centerline. Since the aircraft can be narrower (only one person wide) tandem-seaters are theoretically faster. Tandem seaters are usually more comfortable to ride in, as each occupant gets at least 24 inches of shoulder room rather than sharing 40 inches with their passenger. However, communication between occupants requires strong lungs or an intercom.

Side-by-side aircraft are more civilized. Both occupants can see the instruments and easily bellow into their corider's ear. Since aircraft engines are as wide as two people anyway, the performance cost is often negligible. If you've got a spouse who's not especially thrilled with flying, they're more likely to be comfortable sitting alongside, rather than stuffed in the back of the plane (Fig. 3-7). Twenty years after I gave my mother her first (and only) lightplane ride, she told me she'd been claustrophobic in the back of the Citabria. She couldn't see forward!

One last configuration consideration is the landing gear. The main choice is the basic configuration: conventional (taildragger) or tricycle gear. Conventional gear is more rugged and better suited to rough and short fields. A tailwheel is usually less complex and is subjected to less strain than a nosewheel. This results in lower maintenance costs. Again, conventional geared aircraft are theoretically faster than trigeared types, since tailwheels are less draggy than nosewheels.

Most pilots are more comfortable with tricycle gear. There's no question that trigears are easier to operate and are safer. A taildragger checkout is sometimes difficult to arrange, too. But don't let taildraggers scare you off. You don't have to be Chuck Yeager to taxi and fly a Champ.

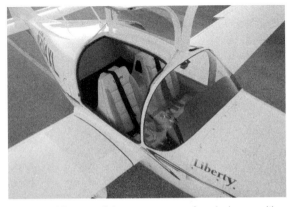

Fig. 3-7. *Most nonflying spouses prefer airplanes with a side-by-side seating arrangement. (Courtesy Liberty Aerospace, Inc.)*

The budget factor

The biggest budget factor when airplane shopping is the age of the aircraft. You'll pay more for a newer airplane. Buying new gives you more trouble-free flying, but buying a used airplane can give you a higher-performing aircraft at lower purchase cost.

How do prices vary among planes of the same age? An old car-racing saying applies: "Speed costs money. How fast can you afford to go?" In actuality, replace "speed" with any of the items in your preferred features list. As the "capability" knob is turned past the "basic airplane" setting, the "cost" meter starts rising.

Things aren't that simple—fortunately. There are probably one or two of your requirements that are inviolate, such as the number of seats the airplane must have to carry your family. If you can accept compromises in some areas, you can likely save some money.

Another major budget factor is the avionics you expect in your airplane. A simple "puddle-jumper" needs just a VFR panel and comm radio (and a transponder, if you're planning on flying anywhere near Class B airspace, unless the airplane lacks an electrical system). A long-legged cross-country machine will need a full IFR panel with dual comms, and serious flyers will want an up-to-date moving-map navigation system.

The difference is tens of thousands of dollars, possibly even surpassing a cool $100,000 for the full-blown, no-holds-barred electronic panels. So chose your mission requirements well.

The big thing to keep in mind is that the operating cost of an airplane is proportional to what it would cost new, not what you paid for it (Fig. 3-8). For instance, you could buy a mid-1950s Beech Bonanza for the cost of a mid-1970s

Fig. 3-8. Surplus Russian-built Antonov AN-2s are occasionally imported to the US, and sometimes sell for about the price of a used Cessna 172. But the operating costs are related to what a new example would cost—the big Russian biplane is a lot more expensive to operate.

Cessna 172. But the ownership cost of that Bonanza will be a lot higher—you'll be paying for more gas being burned in a larger engine, and it'll cost more to maintain the retractable gear and constant-speed prop. Chapter 5 addresses other issues related to buying used aircraft, while Chap. 4 covers the new-aircraft market.

A SAMPLE SELECTION

As of this point, you should know your mission and the performance, capability, and particular features you need. This list should include:

1. Endurance
2. Cruise speed
3. Number of seats/Useful load
4. Seating configuration (if two-seater)
5. Avionics level (Basic VFR, IFR, GPS, and so forth.)
6. Landing gear configuration.
7. Any special performance requirements (short field, off-airport operations, and so on.)

Let's examine how a typical selection process might go. Say, for instance, that your mission is something like: "Two seat side-by-side airplane for fun flying and sightseeing. Cost less than $30,000 to buy, and must be inexpensive to own and operate." A pretty typical first-time-buyer's mission.

The first cut

That $30,000 limit restricts you from new aircraft, so let's assume you're going with a used plane.

Specifying a two-seater further narrows down the field. The "sightseeing" aspect kind of aims us toward a high-wing airplane, eliminating the Piper Tomahawk and the Ercoupe. Side-by-side seating cuts out planes like Citabrias or Aeronca Champs. You're left with planes like the Aeronca Chief, Luscombe, Cessna 120/140/150/152 series, Arctic Tern, and the Taylorcraft (see Table 5-1 in Chap. 5).

You've also specified the need for inexpensive ownership. The cheapest storage option is an outside tiedown which rules out most of the fabric-covered planes. What are we left with? The two-seat Cessnas (except the 120 and early 140s, which had fabric covering on the wings) and the Luscombe. Luscombes are fine airplanes, but they are getting a bit rare.

So, now we've established we're in the market for a nice Cessna 140, 150, or 152 (Fig. 3-9). They're really a good pick for a first-time aircraft. The line has good parts availability, and any A&P will be well familiar with their problems and foibles.

Prices? They vary, but by entering the term "used aircraft for sale" on your favorite Internet search page (Yahoo, Google, and the like.) you'll come up with a range of sites that'll show how the prices run. So, should you just find the best one in your price range and start negotiating? You could work that way. The physical condition and reliability of the aircraft will have the most contribution

Fig. 3-9. *The Cessna 150/152 series is a good low-cost starter aircraft.*

to your satisfaction with ownership. Pick one that's mechanically sound and you'll probably be happy with it.

Still, wouldn't you like the most for your money? The 140/150/152 line was built for over 30 years; you can get the features that fit you the best by picking the right year. For instance, older versions of the Cessna 150, have a higher useful load than newer models.

Let's see how some of the changes might affect your aircraft selection. We'll keep the two-seat Cessnas for an example. Even if you're interested in a larger, more complex aircraft, please follow along. Most aircraft show similar evolution over their design life.

External differences

Figure 3-10 shows the external changes in the 140/150/152 series over the years. The line started out as a rounded-tail conventional-gear aircraft. The 150 series started in 1959, with the major apparent change of the squared-off tail (to match the 172) and tricycle gear. The "tunnel" effect of the fuselage was relieved in 1964 with the addition of the cockpit's rear window ("Omni-vision," Cessna called it).

With the addition of a swept tail, the '66 model was the first of the "modern" 150s. The '71 lengthened the dorsal fin; the '75 added a slightly taller tail. Finally, the tried-and-true Continental O-200 was replaced by the Lycoming O-235, resulting in a different-shaped cowl and a designation change to the Model 152.

Which looks better? You probably said the '66-models on, since the swept-tail birds are more common. Still, did you ever wonder why the tail was swept? A 100-knot airplane didn't need it for speed. It was instigated by the Cessna marketing department.

Did it produce an aerodynamic benefit? To the contrary, the rudder is actually *less* effective than that of the earlier, straight-tail models. The hinge line isn't perpendicular to the relative wind. The tail height was raised 6 inches in 1975, gaining back some of the lost efficiency.

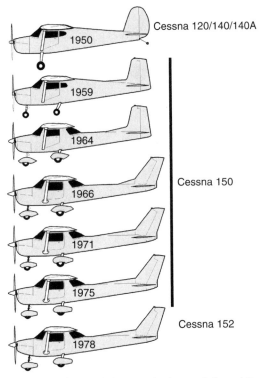

Fig. 3-10. *External changes in the evolution of the Cessna 140 through the Cessna 152.*

Don't be too hard on the Cessna marketers. Sales of the 150 simply exploded with the '66 model. Aerodynamically wasted or not, the swept-tail 150s sold like gold-plated hotcakes.

Let's look back even further, before the rear window. It was also a marketing response, but one, most pilots agree with. That extra Plexiglas area makes the cabin far airier and more pleasant. The straight-back earlier models look almost old-fashioned. Then again, perhaps that's an advantage. One problem with owning a 150: Whenever you fly, people assume you're a student. You walk into a FBO after a flight, and people ask, "Do you need someone to sign your logbook?"

You won't have that problem with a straight-backed 150. It's a poor-person's classic airplane; rare enough to be noticeable, yet as cheap to run as any other 150. The 140 is even better. So if you don't want to look like the crowd, you might set your sights on an earlier model. True of more aircraft than just the Cessna 150 series.

Landing gear

Easy enough to see the difference here, right? The 140 is a taildragger, and all the 150s and 152s are tricycle.

Still, you may not want to lump all the 150s/152s together. There are subtle differences through the line, with several changes. The first came in 1961, when the main gear was moved back 2 inches. Seems the plane had a tendency to fall back on its tail if you fully-loaded the baggage compartment before anyone got in. The gear was changed again 10 years later. The track of the main gear was increased to provide more stability. Lack of stability, of course, is the problem with the Cessna 140's conventional gear. In addition to groundloops, some 140s had a tendency to nose-over as well. Look at the gear legs of a typical 120 or 140, often you'll see a set of aftermarket extenders, as shown in Fig. 3-11. They move the main wheels forward a few inches to reduce this tendency.

Should you reject the 140, just because of the conventional gear? Of course not. All it takes is training, and a little more attention. During the 1950s and 1960s, a number of airplanes transitioned from taildragger to tricycle gear. The Cessna 170 became the immortal 172. The Cessna 180 became the 182. The Piper Pacer begat the Tri-Pacer. Since fewer buyers are interested in taildraggers, their prices sometimes are lower than their trigear equivalent.

Engine choices

Unless you've had a bad experience or two, you probably don't have any biases about engines. One thing to keep in mind: When you go outside the normal Lycoming and Continental arena, you're likely to encounter rare and/or high engine part prices.

Not much of a problem with our target aircraft line. The 140 had an 85 or 90-horse Continental and the 150 the tried-and-true O-200 (100 horsepower). The 152 is essentially a 150 with a Lycoming O-235 substituted for the Continental. And therein lies one problem. The reason Cessna made the swap was so the 152 could run on 100LL

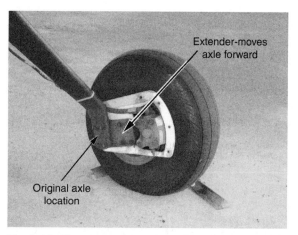

Fig. 3-11. *The noseover tendency of the Cessna 120/140 series brought about these aftermarket extenders. Note how they move the wheel a few inches forward of the gear leg.*

aviation fuel. The Continental was designed for 80 octane; even the lower lead content of 100LL fouled spark plugs or otherwise caused problems. Most of the major fuel refiners were eliminating 80-octane fuel; so, Cessna felt bound to change.

But a few years later, the Experimental Aircraft Association (EAA) ran an extensive set of flight tests and gained approval to use ordinary unleaded auto gas in the Cessna 150. At one swoop, the EAA reduced the direct cost of operating a 150 by almost one-half!

So, if you buy a Cessna 152, you'll be stuck with using high-priced aviation fuel. It is actually possible to modify the engine to gain auto-fuel approval, but a number of high-priced engine parts must be replaced.

Using autofuel in your airplane is not a free ride. If you plan to fly a lot of cross-countries, you'll find 100LL a lot more available on most airports than 80 octane or autofuel. In this case, it'd be better to own an airplane designed to use 100LL. But for the lowest cost of ownership, pick a plane for which Supplemental Type Certificate (STC) approval is available.

Equipment and features

Just because a plane is older doesn't mean it contains out-of-date, obsolete avionics. A 1964 Cherokee probably has had its avionics upgraded since it was new; a 1976 model might still have its original suite. Then again, maybe that '64 model was upgraded in 1972.

There are other features to consider. Let's look back at the 150 series again. The original models had manually-activated flaps. A large lever (like a small car's emergency brake) is mounted between the seats. Pull the bar upward, and the flaps come down. The mechanism had notches at 10, 20, 30, and 40 degrees.

In 1966, Cessna added an electric motor to the flap system and eliminated the lever. An unfortunate decision, in my opinion. The mechanical system was a lot more reliable; in fact, the electric flaps were soon the subject of a recurring Airworthiness Directive (AD). The mechanical flaps are a lot faster to operate, and you can tell the flap position with your right arm rather than having to glance at a panel indicator.

When Cessna transitioned to the 152, they made one more change. They eliminated the 40-degree flap setting. Too many pilots were crashing during go-arounds by forgetting to raise the flaps. This was a pretty common change to the Cessna line about this time, so if you need a short-field bird, you may want to concentrate on earlier models. Cessna made another change to its lines starting around 1978—the conversion to 24-volt electrical systems.

Now, I know the engineering advantages of the 24-volt bus. Wire diameters can be reduced. Starters have a bit more ooomph. Yet a 24-volt system is a 24-carat hassle as far as the private aircraft owner is concerned. Twenty-four volt batteries cost almost double. Ordinary parts such as bulbs and solenoids cost more, and are far less available. You can buy a cheap 12-volt battery charger at any auto parts store; you'll have to get a 24-volt version from an aviation supplier. And if you're ready to fly and the aircraft's battery is dead, you can't jump-start it from your car.

What about age?

We've been talking a lot about buying older airplanes, from the 60s and even the 40s and 50s. Are there dangers involved with planes that are old? What about corrosion? True, you'll have to look out. But corrosion isn't just an old-airplane phenomena. It can happen with relatively recent models, too, depending on where they're based and how they're kept.

Parts can be a problem with older planes. The Continental A-65 engine used in a lot of the small planes of the 50s and 60s has been out of production for over 40 years. Several aftermarket companies have ensured that parts stay available for this and other out-of-production engines, although the prices continue to rise. Similarly, companies have sprung up to supply replacement parts for the airframes as well. Some companies manufacture enough parts to practically build a complete plane, although the price would be a tad high.

The additional maintenance involved with older airplane is often balanced by the lower purchase price.

Where do you find out this stuff?

There's been a lot of information about the Cessna 150/152 series packed into the last couple of pages. Obviously, you'd like a similar level of detail for the type of plane you're interested in.

The Internet to the rescue again. Pick a given aircraft type, then enter it on your favorite web-search page with the words "owners" appended to it, for example, "Cessna 150 owners." You'll not only find web pages run by individual owners, but those of ownership clubs. Entering "Cessna 150 owners," for instance, finds the link for the Cessna 150-152 Owners' Club (*http://www.cessna150-152.com/*). The site has extensive online references, as well as forums where you can discuss the airplane with current owners.

Aviation Consumer magazine is another good source for ownership information. They publish an in-depth analysis of a specific aircraft type in almost every issue. These articles include the differences between models, maintenance problems, safety record, and price history. The magazine is published monthly, and much of their information is available to subscribers online. The magazine's web page is *www.aviation-consumer.com*

The magazine's aircraft analyses have been reprinted *The Aviation Consumer Used Aircraft Guide,* published by McGraw-Hill.

Another good source of information is Bill Clarke's *How to Buy a Single-Engine Airplane (Illustrated Buyer's Guide)*, also published by McGraw-Hill. Clarke discusses a number of aircraft from the owner's point of view, and provides a summary of AD notes on each model.

Books on specific aircraft lines are widely available (Fig. 3-12). There are several on the Cessna and Piper singles, and other volumes on specific models. A search on any online bookstore's web page should reveal what books are available for the airplane you're interested in.

Fig. 3-12. *Your local aviation bookstore will have dozens of books on various popular general aviation aircraft.*

LOOKING AT THE OPTIONS

In Chaps. 4, 5, 6, and 7 we're going to look at some of the basic options in purchasing an aircraft. Chapter 4 will discuss buying a new airplane and Chap. 5 will lead you through the used-airplane buying process.

Chapters 6 and 7 will examine some less-common options. Chapter 6 will explore the ramification of option for *really* used airplanes—antiques and classics. Chapter 7 will look at a special alternative: building your own airplane, or buying a used homebuilt.

CASE STUDY: A CHIEF FOR FUN FLYING

Lynn Berkell had no trouble determining what her "mission" was: "There are pilots who like getting up and looking down on the world... just going up and looking around. That's the group I belong to."

A 1947 Aeronca 11AC Chief seemed perfect for her needs (Fig. 3-13). "It just came up by surprise," she says. A meat-cutter at a local grocery, Berkell was looking for a low-cost plane, and the Aeronca seemed to fit the bill. Airplane ownership wasn't completely new to her. She'd been partners on two earlier airplanes, a Cessna 172 and a Beech Bonanza. And her pilot-husband, Arthur, owns a Mooney.

The Chief was owned by a man at a nearby airport, and the airplane was fairly well-known in the local pilot community. Berkell used some savings and the money from the sale of her previous partnership and bought Chief NC3954E (Antique and Classic airplanes can use the old-time "NC" prefix). She bought the plane in 1999, paying less than $20,000.

The airplane wasn't much different from the day it had rolled out of the Aeronca factory (Fig. 3-14). Its had no electrical system, which means it was not required to

Fig. 3-13. *Lynn Berkell's Aeronca 11BC Chief.*

carry a transponder unless she actually wants to enter Class B or C airspace. There was no starter, either. "I had learned to hand-prop when a friend let me fly his Champ," says Berkell. The Chief has a glider hook installed in the tail, with a cockpit release. This allows the pilot to tie a rope to the tail, start the engine, climb into the cockpit, and release the rope.

Ownership experience

Ask Lynn Berkell what her maintenance costs on the Chief are, and she responds, "About a million dollars." A bit of an exaggeration, yes, but she had some rather unfortunate experiences, early on.

Fig. 3-14. *Berkell's panel is simple elegance. Since the plane lacks an electrical system, the panel doesn't include common items like switches, fuses, lights, ammeters, radios, or a transponder.*

After putting about 30 hours on her new airplane, one of the cylinders lost compression. Since the engine was going to be torn down anyway, she opted to upgrade the engine from 65 to 75 horsepower, using one of the common Continental STCs.

She turned the engine over to a local mechanic, and waited. After an entire year, the man had only completed half the rebuild. She demanded the engine back and passed it to another mechanic. This one took eight months, and installed the wrong cylinder rings. They disintegrated in 20 hours, and the mechanic refused to stand behind his work. So she had to pay for *another* rebuild.

Her annuals cost about $1500, which includes the repair of the typical problems that come up within a year. After her bad experiences with the engine rebuild, she asked a local A&P she trusted (but who didn't have the time to take her as a customer) to recommend an A&P/IA for her annuals. He suggested a man at a nearby airport, and Berkell is very pleased. "He's a great mechanic," she says. "He calls you immediately, and he's done when he says he's going to be done."

Full-coverage insurance for the Chief costs $1200. She had to make a claim several years back. She had stopped at a nearby airport to refuel. The airport previously had a chain wrapped around a post to secure planes for hand-propping, but she discovered it had been removed. She used a piece of rope, but it gave way and the plane came free and rammed an airport structure.

She flies about 35 hours a year, mostly local pleasure flights, with one ~150 mile trip to an antique airplane fly-in. Since she works on weekends, she doesn't have the opportunity to fly to many of fly-ins or go on group flights with friends. Also, her hand-prop accident (and the subsequent FAA attention) makes her uncomfortable with stopping at nearby airports unless she knows someone there who can prop her airplane.

Berkell and her husband have a home on an airpark, and Chief Fifty-Four Echo resides in their own hangar with his Mooney. She enjoys the airpark life. "I was surprised at how nice everyone is," says Berkell. "It's like a club."

A summary of Berkell's ownership is presented in Fig. 3-15.

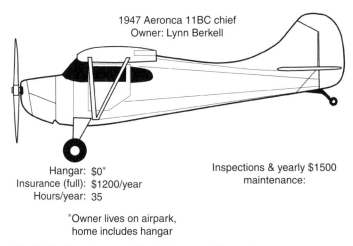

1947 Aeronca 11BC chief
Owner: Lynn Berkell

Hangar: $0*
Insurance (full): $1200/year
Hours/year: 35

Inspections & yearly $1500
maintenance:

*Owner lives on airpark,
home includes hangar

Fig. 3-15. *A summary of Berkell's ownership costs.*

Advice

Lynn Berkell enjoys not having to worry about the condition of the accessories like the intercom, and the ability to just leave her headsets in the airplane. "If you're working on a rating, you don't have to worry about scheduling."

Should you buy an airplane instead of renting? Lynn Berkell says, "Do it! By all means, do it!"

4

That New-Plane Smell

One of the largest budget factors on an aircraft purchase rides on a single decision: Are you going to buy a new airplane, or a used one?

When it gets right down to it, who wants a used *anything*? A used car, a used house, a used dog, a used toothbrush. Given a choice, we'd all prefer to buy everything brand-new.

There's probably not a pilot alive who hasn't wanted a shiny new airplane. One they can treat right from the beginning. One with a modern panel, equipped from the factory with up-to-date avionics. A beautiful, clean interior rather than that scruffy 60s fabric (Fig. 4-1).

And, of course, you've got a brand-new airplane without the thousands of hours of wear the average General Aviation aircraft has. From my personal perspective, the thing that irritates me the most about owning older airplanes is not the major wear items. It's the *little* stuff going bad. Like breaking a door handle, and realizing that the little five-buck part has been out of production for 30 years. And that I would be in for a series of phone calls to aircraft scrapyards to try find a replacement. With a new plane—it's just a matter of a call to the factory for a replacement.

General Aviation went through a considerable depression in the 1980s and 1990s. Both Cessna and Piper stopped producing small single-engine aircraft and several other companies closed. New General Aviation aircraft became quite scarce. Some years during this period, more light aircraft were built in garages (for example, homebuilts) than on factory floors. In 1980, over 10,000 new General Aviation aircraft received FAA registration. Fifteen years later, registrations had dropped by 95 percent!

Things have bounced back since then. Companies like Diamond, Lancair, and Cirrus have brought modern composite construction to the General Aviation market, and Cessna (Fig. 4-2) and New Piper are building updated versions of their all-metal machines. Also, a new certification category, special Light Sport Aircraft (LSA), is starting to produce a whole new breed of production aircraft.

Years ago, thousands of small aircraft were built every year, by dozens of companies. Those days are long past, and with them, the variety that spiced

Fig. 4-1. *One can certainly update the interior and panel of an older airplane, but they can't really match the look and feel of a brand new cockpit. (Courtesy Columbia Aircraft.)*

up the flight lines. But the revitalization of General Aviation brought about some new thinking. As Table 4-1 illustrates, traditional aircraft still have a large share of the market, but twenty-first century designs have taken a strong foothold.

Fig. 4-2. *Cessna returned to production in the late 1990s, eventually bringing the Cessna 206 StationAir back onto the market. (Courtesy Cessna Aircraft Company.)*

Table 4-1. New standard category light aircraft

Manufacturer	Model	Engine Make	HP	Gross (lbs)	Cruise (kts)	Range (NM)	Seats/Config	Gear	Price($) 2005
AMD	Alarus	Lycoming	115	1692	100	575	2 SBS	Tri.	119,900
American Champion									
	Adventure	Lycoming	160	1750	117	365	2 Tan	Conv.	100,900
	Aurora	Lycoming	115	1750	100	500	2 Tan	Conv.	89,900
	Explorer	Lycoming	160	1800	111	350	2 Tan	Conv.	103,900
	Super Decathlon	Lycoming	180	1950	111	475	2 Tan	Conv.	131,900
	Scout	Lycoming	180	2150	113	305	2 Tan	Conv.	123,900
Aviat	Husky A-1B	Lycoming	180	2000	122	565	2 Tan	Conv.	149,700
	Husky Pup	Lycoming	160	2000			2 Tan	Conv.	119,579
	S-2C Pitts	Lycoming	260	1700	148	250	2 Tan	Conv.	194,295
Beech	A36 Bonanza	Continental	300	3650	176	700	6	Tri-Retract	650,040
Cessna	Skyhawk	Lycoming	160	2450	122	580	4	Tri.	164,250
	Skyhawk SP	Lycoming	180	2550	126	525	4	Tri.	171,250
	Skylane	Lycoming	230	3100	143	840	4	Tri.	258,500
	Turbo Skylane	Lycoming	235	3100	157	590	4	Tri.	286,500
	Stationair	Lycoming	300	3600	143	605	6	Tri.	367,500
	Turbo Stationair	Lycoming	310	3600	165	700	6	Tri.	400,500
Cirrus	SRV-G2	Continental	200	3000	150	630	4	Tri.	189,900
	SR20-G2	Continental	200	3000	157	740	4	Tri.	236,800
	SR22-G2	Continental	310	3400	183	700	4	Tri.	334,700

Make	Model	Engine					Seats	Gear	Price
Classic	Waco YMF Super	Jacobs	275	2950	109	375	2 Tan	Conv.	254,500
Commander	115	Lycoming	260	3260	161	850	4	Tri-Retract	450,500
	115TC	Lycoming	270	3305	187	670	4	Tri-Retract	498,000
Diamond	DA20-C1 Eclipse	Continental	125	1764	139	480	2 SBS	Tri.	146,990
	DA40-180 Diamond Star	Lycoming	180	2535	143	600	4	Tri.	206,430
Interstate	Arctic Tern	Lycoming	160	2150	109		2 SBS	Conv.	147,100
Lancair Certified Aircraft	Colombia 300	Continental	310	3400	190	960	4	Tri.	285,500
	Colombia 400	Continental	310	3600	230	1090	4	Tri.	329,500
Liberty Aerospace	Liberty XL2	Continental	125	1653	131	435	2 SBS	Tri.	139,500
Maule	M-4-180V	Lycoming	180	2350	109	825	2	Conv.	95,999
	MX-7-180B	Lycoming	180	2500	126	980	4-5	Conv.	130,850
	MXT-7-180B	Lycoming	180	2500	123	825	4-5	Tri	142,120
	M-7-235B	Lycoming	235	2500	139	750	5	Conv.	147,200
	MT-7-235	Lycoming	235	2500	138	740	4-5	Tri	168,100
Mooney	Ovation	Continental	280	3368	191	1565	4	Tri-Retract	299,000
	Ovation2 DX	Continental	280	3368	191	1565	4	Tri-Retract	397,750
	Ovation2 GX	Continental	280	3368	191	1800	4	Tri-Retract	427,450
	Bravo DX	Lycoming	270	3368	213	980	4	Tri-Retract	448,750
	Bravo GX	Lycoming	270	3368	213	1150	4	Tri-Retract	472,450

(*Continued*)

Table 4-1. New standard category light aircraft (*Continued*)

Manufacturer	Model	Engine Make	HP	Gross (lbs)	Cruise (kts)	Range (NM)	Seats/Config	Gear	Price($) 2005
New Piper	Warrior	Lycoming	160	2440	117	515	4	Tri.	161,000
	Archer	Lycoming	180	2550	129	445	4	Tri.	188,900
	Arrow	Lycoming	200	2750	139	875	4	Tri-Retract	249,700
	6X	Lycoming	300	3600	148	800	6	Tri.	365,000
	6XT	Lycoming	300	3600	161	930	6	Tri.	
	Saratoga II HP	Lycoming	300	3600	165	860	6	Tri-Retract	425,700
	Saratoga II TC	Lycoming	300	3600	184	960	6	Tri-Retract	456,100
	Malibu Mirage	Lycoming	350	4340	222	1350	6	Tri-Retract	539,075
Sky Arrow	650 TCNS	Rotax	100	1433	90	370	2 Tan	Tri	84,000
Socata	TB9 Tampico	Lycoming	160	2337	115	440	4	Tri	190,390
	TB10 Tobago GT	Lycoming	180	2530	127	650	4–5	Tri	223,980
	TB200 Tobago GT	Lycoming	200	2535	130	650	4–5	Tri	246,200
	TB20 Trinidad	Lycoming	250	3086	164	760	4–5	Tri-Retract	360,200
	TB2T Trinidad GT	Lycoming	250	3080	190	750	4–5	Tri-Retract	405,200
Symphony	160	Lycoming	160	2150	129		2 SBS	Tri	139,900

SHOPPING

There are two basic ways new planes are sold: Either from a local dealer, or direct from the factory. Smaller companies tend to handle their sales directly, while larger ones work through dealers.

The nominal value of a dealer is that they make it easier to examine a demonstrator aircraft and take a test flight—you don't have to travel to the factory. However, the day of a dealer at every airport is long past. Living in Seattle, two major brands are handled within 25 miles, but the closest dealer for another major brand is a thousand miles away. While most manufacturers attend the major airshows (like Oshkosh and Sun-N-Fun), it can be difficult to receive more than a hurried demo flight at these events (Fig. 4-3).

Even if one isn't located nearby, there can be some advantages to working through a dealer. A dealer is more likely to present you with options—options for financing, equipment, or even be able to suggest an alternate approach, such as a late-model used plane or a fractional ownership arrangement. On the other hand, they might push you toward a deal that maximizes their own profit, rather than what's best for you.

You'd think that with hundreds of thousands of dollars on the line, aircraft sales representatives would be quite attentive and helpful. Surprisingly, this is not always the case—one man I spoke to felt that he was given short shrift by the salesman for one company. The bigger companies usually sell large corporate-class aircraft and some sales personnel may be less interested in customers for the small single-engine planes.

Fig. 4-3. *Companies bring aircraft like the Liberty XL2 to major fly-ins, but you'll have a better chance for a good demonstration ride if you can contact a local dealer. (Courtesy Liberty Aerospace, Inc.)*

FINANCING

Aircraft manufacturers know their products are expensive. If they don't offer their own finance plans, they or their dealers should be able to put you in contact with a number of banks and finance companies. For example, Cirrus Aircraft offers a variety of financing options—a 3-year adjustable rate loan, a 7-year loan with a balloon payment, even a 20-year fixed rate loan like a home mortgage.

You may be able to find a better loan on your own, but in any case, most institutions will accept the plane itself as collateral.

WARRANTIES

One of the big advantages the new-airplane makers can offer is a full warranty on their products. Just like automobile warranties, these are based on both calendar months and total operating time. Checking around, I found warranties ranging from 12 months and 100 flight hours to 36 months and 1000 flight hours (Fig. 4-4).

Sure, nothing major will go wrong. But new *anything*—cars, houses, and airplanes—suffer teething problems. Warranties get the little stuff fixed at no cost to you.

There are two basic systems for handling warranty work. The larger companies have networks of service centers. These are generally existing Fixed Base Operators (FBOs). The system works like car warranties—bring your plane to it, and if the problem is covered by the warranty, there's no charge.

Fig. 4-4. *Buyers of new planes like this Diamond Eclipse gain some peace of mind due to the factory warranties on the airframe, engine, and avionics. (Courtesy Diamond Aircraft.)*

The smaller companies don't have the service networks. Generally, you'll be required to pay your local A&P to do the work, and file with the company for reimbursement.

The first system works best if you've got an approved service center nearby. Otherwise, it's a bit of a hassle to have to fly a cross-country for service. Keep in mind that some major components, such as the engine and avionics, are *not* covered by the aircraft warranty. The companies that make these units issue their own warranties—and their warranty period may be shorter than aircraft warranty. There can also be some finger-pointing problems. The transponder company might blame intermittent operation on the aircraft power supply, but the aircraft manufacturer might claim there's nothing wrong with the electrical system.

Also, these vendors usually require service to be performed by one of their own authorized centers. Like the service centers for the aircraft, it's not a problem if there's an approved shop nearby. But it's a hassle if there isn't—and especially if there isn't an approved engine or avionics vendor on the same field as the aircraft service center.

DRAWBACKS

The major drawback to buying a new airplane is depreciation. Like automobiles, the value of a new plane starts dropping as soon as you accept delivery. A 1975 Mooney, if you keep it maintained, can probably be sold five years later for something fairly close to what you paid for it. But you have to accept a pretty heavy loss in the first few years after taking possession of a factory-fresh airplane.

There are a lot of new companies entering the new-aircraft market lately. Keep in mind that not all will stay in business. Owning an "orphan" design can be a nightmare, since new replacement parts aren't being made and there aren't very many examples in the boneyards to supply the used-parts networks.

CASE STUDY: A NEW CIRRUS

Patrick Flynn's first plane was a Grumman Tiger—but all it did was whet his appetite. "I needed something with longer legs and more altitude," says the Seattle-area executive. More speed was also a key point. He and his wife, Kathleen, have family all along the West coast, and after 18 months with the Tiger, they decided to buy something that would get them there faster (Fig. 4-5).

They looked at the planes on the market, and eventually decided on the Cirrus SR-22. The key point: Safety. Patrick liked the fixed gear of the SR-22 vs. the retractable gear of the Bonanza. Less complex, fewer things to go wrong, no chance of landing gear-up by mistake. The Cirrus also features a single-lever propeller control that eliminates the need to jockey individual throttle and the propeller controls. "You lose some engine control, but you make up for it in simplicity," he explains.

All attractive to a pilot, but another safety feature was key: "To my wife," remembers Patrick Flynn, "Having the chute was a big selling point." All Cirrus aircraft include the Cirrus Airframe Parachute System (CAPS), an aircraft recovery

Fig. 4-5. *Kathleen and Patrick Flynn had previously owned a Grumman Tiger, but wanted something faster.*

parachute triggered by a single cockpit handle. "Flying was never my ideal activity," recalls Kathleen. But if something happened to her husband in flight, she wanted to be able to bring them down safely. Talking to Cirrus owners, Patrick discovered this was a common theme among nonpilot spouses.

Coupled with the aircraft's performance and comfort, the Flynns were sold. "The sales process was easy," says Patrick. "Much like buying a car, except the tag had more zeros!" They sold some stock options for their down payment, and financed the rest of the $385,000 purchase price through Cirrus. The down payment was sufficiently high to earn them an additional $20,000 discount.

One thing took them by surprise: "I thought we'd have a six-month wait for delivery," says Flynn. "But they called us up and asked, What day would you like to pick it up?" The Flynns traveled to Cirrus headquarters in Duluth, Minnesota, signed for Cirrus N6099Z (Fig. 4-6), and flew it back to Seattle.

They did receive an unpleasant surprise when they returned to Washington state—they were required to pay over $30,000 in sales tax on the aircraft.

Ownership experience

When he picked up his new plane, Patrick Flynn had just 300 hours logged, but has almost doubled that in the 18 months since he bought the plane. His ownership experience has been overwhelmingly positive—only a few minor faults, all corrected under warranty.

One failure was a good thing, in a way. The nosewheel fairing cracked, and the Flynns discovered a replacement fairing cost $1200. While the warranty covered the replacement, the Flynns bought a cover to lessen the chance of damage once the two-year warranty was past.

The Cirrus' avionics did take some time getting used to. Like most pilots, Patrick Flynn learned to fly using the standard analog instruments. The Cirrus replaces these "steam gauges" with an integrated glass cockpit. The panel features

Fig. 4-6. *A large factor in the Flynn's selection of the SR-22 was the standard Cirrus Airframe Parachute System.*

two bright active-matrix liquid crystal displays, which display all flight and operations information. Primary flight data like airspeed and altitude are now displayed on a sliding scale, like modern military and commercial aircraft.

"It took about ten hours to get used to everything," Flynn says. "But now, the situation awareness is unbelievable. It would be tough to go back." The panel includes backup mechanical flight instruments. He has all the aircraft maintenance performed at the nearest Cirrus service center. They charge $1600, plus parts, for the annual inspections. His first inspection ran $2200, but Flynn expects the next to be a few hundred dollars less on the next one.

As a lower-time pilot, insurance was initially fairly expensive. The first-year premiums were $9500. But now, with almost 300 hours in type, the annual premiums have dropped by over $4000.

Patrick Flynn has some modifications in mind. He's looking at a particular navigation system upgrade, but is going to let the technology mature a bit. An earlier-generation real-time satellite weather data link didn't work out, so he's considering a change to another service. The Flynns currently use a portable oxygen system for their high-altitude flying, but Patrick is planning on having a built-in system installed to reduce the clutter from the hoses.

A summary of the Flynn's ownership presented in Fig. 4-7.

Advice

"We wanted to go fast and far," says Patrick Flynn. Cirrus N6099Z meets their needs nicely. They fly from Seattle to California's Napa Valley in just 4 hours—less time than flying by airlines would take, by the time one deals with airport security.

Patrick and Kathleen are very happy with their Cirrus, and buying a brand-new bird minimized the risks involved in used airplanes. "I know exactly what its history is," says Patrick. With a brand-new airplane, he doesn't have to worry that a

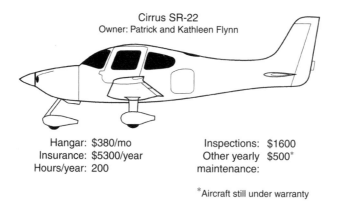

Cirrus SR-22
Owner: Patrick and Kathleen Flynn

Hangar: $380/mo
Insurance: $5300/year
Hours/year: 200

Inspections: $1600
Other yearly $500*
maintenance:

*Aircraft still under warranty

Fig. 4-7. *A summary of the Flynn's ownership costs.*

previous owner deferred maintenance or hid damage. Owning the airplane also brings him more confidence when he flies. "I'm more involved with the airplane," he says. "I know what the engine should sound like."

When problems do arise, he has a factory-support structure to rely on. "The Cirrus service center is a big part of making things go smoothly," says Flynn. "I really value the spinner-to-tail warranty coverage." The CAPS was a big factor in selecting the Cirrus. Patrick says, "Pilot incapacitation is the biggest fear of spouses. Now they have an option."

Kathleen Flynn has shed much of her nervousness. "I felt more confident once I took my first flying lesson." Her advice for nervous spouses: "The more you know, the less you fear."

When it comes time for shopping for an airplane, Patrick has clear advice: "Fly them, figure out your mission, and buy all the airplane you can. If you scrimp, you won't be happy."

LIGHT SPORT AIRCRAFT

In the year 2004, a revolution occurred in General Aviation. The FAA instituted two programs that heralded a major revolution in the American regulatory environment: The Sport Pilot license, and the new Light Sport Aircraft (LSA) category. These new programs greatly expand the horizons for General Aviation.

There are four major aspects of the new system: The Light Sport definition, the Sport Pilot license, the Special LSA certification category, and the Experimental LSA.

The light sport definition

The centerpiece of the new rules is a legal definition of a new type of aircraft—Light Sport. These are airplanes with no more than a single engine, with a gross weight of 1320 pounds, no more than two seats, fixed gear, fixed-pitch (or ground-adjustable) propeller, a stall speed less than 52 mph, and a maximum level-flight speed of 138 mph. A summary of these characteristics is shown in Fig. 4-8.

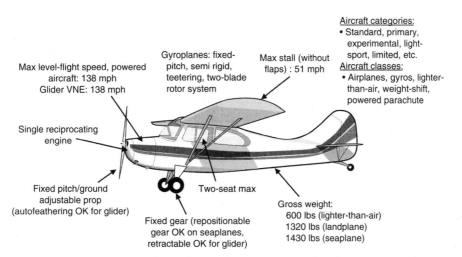

Fig. 4-8. *Required characteristics of an airplane meeting the Light Sport Aircraft definition.*

An airplane does not have to be certified in any specific category to meet the Light Sport definition. Classic production planes like Piper Cubs and Aeronca Champs qualify, as do many homebuilts. Of course, aircraft that are certified in either of the two new categories (Special Light Sport and Experimental Light Sport) also qualify.

The sport pilot license

The sport pilot license was developed in conjunction with this new aircraft definition. It recognizes that pilots flying simpler aircraft don't need much of the training included in the standard private pilot curriculum. A person may earn a sport pilot license in as little as 10 hours total time. Persons holding recreational or private licenses can exercise sport pilot privileges without needing additional training or passing any tests, as long as they comply with the medical and other limitations of the sport pilot rules.

The new license includes one greatly-anticipated feature: Sport pilots are not required to obtain an FAA medical. The FAA accepts that the possession of a valid state driver's license demonstrates that the holder has the minimum level of health necessary to fly these simple aircraft. As long as their driver's license is valid, the pilot is allowed to assess his or her health and determine if any medical condition precludes safe flight.

Unfortunately, there's one big exception: If a person had previously possessed an FAA medical and their most recent medical had been suspended or revoked, that person will not be able to use a driver's license in lieu of a medical. The same situation holds true if the person's most recent "Authorization for Special Issuance of a Medical Certificate" had been withdrawn. Currently, negotiations are underway to hopefully develop procedures to allow affected individuals to still operate as sport pilots.

Sport pilot students start out the same way as any other student, except they are not required to obtain a Class III medical. Aeronautical experience required for the license is pretty basic. Between 10 and 20 hours of flight time are necessary, depending upon the class of aircraft (glider, powered parachute, airplane, and so forth.). In addition to flight training, students must either receive ground training from an instructor or complete a home-study course on the usual topics, such as aerodynamics weather and regulations. Both, a knowledge and a practical test (including both oral and flight portions) are required to earn the sport pilot license.

Like other license-holders, sport pilots must undergo a flight review every two years. Time flown as a sport pilot counts toward the aeronautical experience requirements for higher licenses. All those flying under sport pilot privileges (including private pilots using a driver's license in lieu of a third-class medical) are limited to aircraft that meet the FAA's LSA definition. However, the aircraft does not have to be licensed in any particular category. Sport pilots can fly standard, primary, experimental, limited, and just about any other category of aircraft. Any aircraft that meets the standard can be flown by a sport pilot (Fig. 4-9).

Sport pilots are limited to day VFR, at altitudes of 10,000 feet MSL or below. They must fly with visual contact to the ground—no "VFR on Top." They cannot fly an aircraft for a sales demonstration or as part of a business, nor can they tow any other aircraft. They cannot fly within Class B, C, or D airspace, unless they receive additional training and the appropriate endorsement. Even with this endorsement, a sport pilot is not allowed to land or take off from airports that require at least a private pilot license (these airports are identified in Part 91 Appendix D, Section 4).

Finally, the license is not valid outside the United States, unless that country grants specific permission. A comparison of pilot privileges is presented in Fig. 4-10.

Fig. 4-9. *Any aircraft meeting the Light Sport Aircraft definition can be flown on sport pilot privileges. The aircraft can be in any certification category—normal, aerobatic, utility, experimental, or the new Special Light Sport Aircraft category. This homebuilt Murphy Renegade qualifies.*

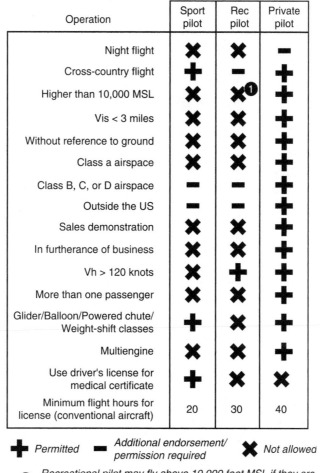

Operation	Sport pilot	Rec pilot	Private pilot
Night flight	✖	✖	▬
Cross-country flight	✚	▬	✚
Higher than 10,000 MSL	✖	✖❶	✚
Vis < 3 miles	✖	✖	✚
Without reference to ground	✖	✖	✚
Class a airspace	✖	✖	✚
Class B, C, or D airspace	▬	▬	✚
Outside the US	▬	▬	✚
Sales demonstration	✖	✖	✚
In furtherance of business	✖	✖	✚
Vh > 120 knots	✖	✚	✚
More than one passenger	✖	✖	✚
Glider/Balloon/Powered chute/ Weight-shift classes	✚	✖	✚
Multiengine	✖	✖	✚
Use driver's license for medical certificate	✚	✖	✖
Minimum flight hours for license (conventional aircraft)	20	30	40

✚ Permitted ▬ Additional endorsement/ permission required ✖ Not allowed

❶ Recreational pilot may fly above 10,000 feet MSL if they are no higher than 2,000 feet AGL

Fig. 4-10. *A summary of pilot privileges.*

Special light sport aircraft

The LSA definition was the basis for a new aircraft certification category—Special Light Sport Aircraft (SLSA).

The "special" denotes that the aircraft do not receive Standard certification, in other words, that they have not been either designed to nor evaluated against the FAA's regulations governing aircraft, engine, and component design, testing, performance, handling, and durability. To this extent, SLSA category is similar to the experimental category.

However, unlike experimentals, SLSA manufacturers are still governed by design standards. Instead of the traditional FARs, though, the design of these planes

Fig. 4-11. *Many light aircraft built ready-to-fly in Europe, such as Aerostar Festival, are being certified as Special Light Sport Aircraft for sale in the United States. (Courtesy Belle Aire Aviation/www.lightsportflying.com)*

is governed by guidelines developed *within the aircraft industry* (Fig. 4-11). The FAA does not dictate the limit loads for airframes or the requirements for SLSA engine certification; the *consensus standards* are developed jointly within the aviation industry and published by American Society for Testing and Materials (ASTM) International, a standardizations organization. (ASTM started as the "American Society for Testing and Materials," but it's now an international organization.)

These new Standards are the equivalent of 14CFR Parts 21 (Certification Procedures for Products and Parts), 23 (Airworthiness Standards: Normal Category), 27 (Airworthiness Standards: Rotorcraft), and 33 (Airworthiness Standards: Engines). My commercial copy of these regulations is over 300 pages long. The ASTM Standards book (Fig. 4-12) is 80 pages, and includes some elements (such as, powered parachutes) that the FARs don't cover.

The consensus standards establish design, manufacturing, quality control, and testing criteria. They also specify the required level of documentation, such as the airplane flight manual and maintenance procedures, and how aircraft companies must track safety-of-flight issues and how the company will ensure that SLSAs in the field continue to conform to the consensus standards. Like homebuilts, they are required to carry a placard indicating that they do not meet Standard airworthiness requirements.

When a manufacturer is ready to produce a normal category airplane, they submit paperwork showing their analyses and the results of their flight test program to the FAA. There, government specialists examine the data to verify that the new airplane does, indeed, meet Part 23 requirements. Government pilots evaluate the aircraft.

The LSA process is quite a bit simpler. The manufacturer submits the aircraft's operating instructions, maintenance and inspection procedures, flight training supplement data, and an affidavit that the airplane meets the LSA consensus standard.

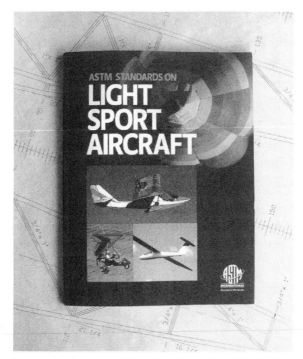

Fig. 4-12. *Certification requirements for the new Special Light Sport Aircraft category are contained in consensus standards developed within the industry, instead of by FAA direction. ASTM International publishes the standards.*

The prototype aircraft undergoes a safety inspection by the FAA, but otherwise, there is no FAA verification of the manufacturer's analysis or test information. The manufacturer merely agrees to allow the FAA unrestricted access to its facilities if the government wishes to verify that the required information exists.

Maintenance and modifications

Once an SLSA is certified, it is treated almost identically to a standard category plane. Special Light Sport Aircraft (SLSAs) can be leased or rented. They require an annual inspection, and, for those planes operated commercially, 100-hour inspections.

Like standard category aircraft, SLSA owners may perform certain preventive maintenance tasks (discussed in a later chapter). All other work, including major maintenance, required inspections, and repairs, must be performed by either an A&P mechanic or by an individual holding the new Light Sport-Maintenance Repairman (LS-M) Certificate. The LS-M certificate can be earned after an 80 to 120-hour course. It allows the holder to perform only those procedures which are spelled out in the manufacturer's maintenance manuals, and in which he or she has received the necessary training.

As part of the consensus standard, SLSA manufacturers specify the allowable replacement parts and any allowed alterations. If unauthorized parts are installed or alterations are performed other than those approved by the manufacturer, the aircraft is considered in violation of the consensus standard and is not airworthy. Technically, this applies to everything. One can change radio types in a standard-category airplane with the FAA's field approval process, but an SLSA *must* remain in the manufacturer-specified configuration.

SLSAs do not fall within the FAA's conventional Airworthiness Directive (AD) system. However, the same level of authority is vested in the manufacturers of SLSAs. They are required to monitor the reliability of their products, and, if necessary, issue safety directives. Compliance is mandatory, though an owner can appeal. Also, if an FAA-certified part is used on the aircraft, it is subject to ADs.

Note that these rules only apply to aircraft that are *certified* as Special Light Sport aircraft. Just because an airplane meets the light sport definition does not affect how it must be maintained or who can maintain it. A normal category airplane like a Champ must still be annualed by an A&P with an Inspection Authorization (IA) rating, even though a sport pilot can legally fly it.

Experimental light sport aircraft

The experimental light sport category was created to allow kit versions of Light Sport Aircraft. Builders of traditional homebuilt aircraft are required to perform the majority of the construction—this is referred to as the "51 percent rule," and is explained more thoroughly in Chap. 7.

However, Experimental Light Sport Aircraft (ELSA) kit buyers are not limited by the "51 percent rule." ELSA kit manufacturers are free to sell their products at whatever degree of completion they wish, from 0 (plans-built) to 99 percent. Of course, there's a catch. To qualify as an ELSA, an aircraft design must first be certified as a Special Light Sport Aircraft. In other words, the manufacturer must design and manufacture a prototype to the consensus standard, and develop all the flight and maintenance instructions the standards require. Once they receive SLSA certification on this prototype, the manufacturer can build ready-to-fly aircraft, ELSA kits, or both.

The decision

Is a LSA the right pick for you? If your mission is the sort that can fit within the regulatory limitations, a SLSA is an excellent choice. They do promise that "new plane smell" for less money. Prices for the first few designs seem to be running around the $50,000–70,000 range.

But keep the limitations in mind. An SLSA cannot fly at night, and maximum-level-flight speed limit will ensure the cruise speed is fairly low. Only one passenger can be carried, and the gross weight limits will restrict baggage capacity. For those looking for a small fun-flying aircraft, an SLSA is a good option.

5

Used Aircraft

Everybody wants a new airplane—but nobody wants to pay for one. It's ironic, really. About 30 years ago, I was fresh out of college, earning a magnificent $1,000 a month as an Air Force second lieutenant. At the time, a year's worth of my Air Force pay could have bought me a brand-new Cessna 172. I left the service years ago for a career in the aerospace field. I've done well, and am now a very senior engineer. My current salary is such that my second-lieutenant self 30 years ago would hardly believe it. Back then, I had a cramped one-bedroom apartment, and today I've got a spacious home overlooking an airport. But, you know—a year's worth of my salary now wouldn't come near to buying me a brand-new Cessna 172.

Happily, there are used airplanes. Assuming a plane has only minimal involvement with hamhanded students, the simple act of flight is not that stressful (Fig. 5-1). It's not too difficult to maintain or replace the engines when necessary. The airframes themselves can fly for thousands of hours without any undue wear.

Buying used airplanes is pretty much the norm, among the pilot community. It's the only way that the "average Joe" can afford his own airplane.

THE DISTORTED MARKET

Most of us have bought used cars at some point in our lives. Unfortunately, the used airplane market is a lot different from the used car market.

The average General Aviation aircraft in the United States is over 30 years old. The recession of the 1980s and 1990s temporarily knocked both Cessna and Piper out of the General Aviation market. They've since returned, but many of the smaller manufacturers have been in and out of business since then. This drop in production is the main reason for the high average age of used aircraft, and is the direct cause of a significant distortion in the used airplane market.

Figure 5-2 illustrates the problem. It plots the number of aircraft on the United States aircraft registration database in January 2005 vs. the year the aircraft were manufactured. Most aircraft in the registry were built during three periods—the immediate post-World War II period, the mid-60s, and the late 70s.

Fig. 5-1. *Merely flying is not that stressful on airframes. With care, used planes like this Mooney can safely fly for decades.*

If you have your heart set on a new BMW but can't afford it, you can look at older and older used Beemers until you find the vintage that matches your budget. Not so on used aircraft—only one out of six airplanes in the United States is less than 15 years old. The limited number of newer used aircraft keeps their prices up; you don't really have a range of choices. You might be able to find a good deal in a late-model used plane, but otherwise, your options are either to buy new or select from the thousands of older used airplanes on the market.

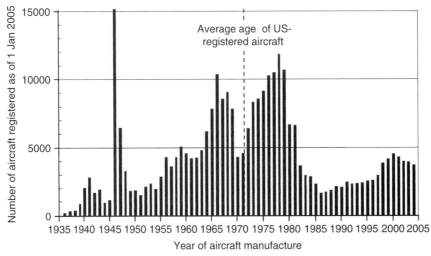

Fig. 5-2. *Year of manufacture for planes listed on the January 2005 FAA registration rolls.*

THE GOOD NEWS

The good news? There's still a vast number of used airplanes.

Consider: There are about 700,000 people with pilot license in the United States. There are about 325,000 registered airplanes. That's only about two pilots per registered aircraft. With so many airplanes, you have a tremendous range of features, performance, and capabilities available in a wide cost range. As long as your requirements don't stray to the extremes of the envelope, you can probably find the kind of airplane you want at a reasonable cost.

And it won't even necessarily be different types of airplanes. Take the lowly Cessna 172, for instance. Its initial production series was from 1957 until 1984. Easy enough, you might think. All you have to do is pick the appropriate year where the average prices correspond with the money you have available.

However, it isn't *quite* as simple as that. Cessna didn't stand still during 27 years of production. The 172 switched from a Continental to a Lycoming in 1968; and to a slightly bigger Lycoming 10 years later. The gross weight was gradually increased from 2200 to 2400 pounds. And the rate of climb started at 660 FPM, rose to 770 by the late 70s, and had dropped back down to 680 by the time production ceased. The XP version offered better performance in the same basic package (Fig. 5-3).

But at least the 172 was built by a single company, under a single name. You're probably familiar with the lowly Ercoupe, a funky-looking twin-tailed cutie from the 40s that offers surprising performance on a 90-horse engine. If you were in the market for one of these, you'd be looking at plane manufactured by companies like Ercoupe, Forney, Alon, and Mooney. By the time Mooney got hold of it, the twin tail was gone and the nameplate said "Mooney Cadet."

So if you're in the market for an airplane, you've got your job cut out for you. But hey . . . picking out what kind of plane to buy is the fun part, remember? We'll discuss some of the issues involved in selection through the rest of this chapter. To get you started, Table 5-1 contains specifications and information for a number of single-engine aircraft.

Fig. 5-3. *The Cessna Skyhawk XP was an upgraded version of the classic Cessna 172. (Courtesy Cessna Aircraft Company.)*

Table 5-1. Used aircraft specifications

Manufacturer	Model	Year	Engine Make	HP	Gross	Useful Load (lbs)	Cruise (kts)	Stall (kts)	Range (nm)	Seats/Config.	Gear
Aeronca	11CC Super Chief	49	Continental	65	1350	530	83	39	365	2 SBS	Conv.
(Bellanca)	7AC Champ	50	Continental	90	1450	560	87	38	304	2 Tan	Conv.
(Bellanca)	7ACA Champ	72	Franklin	65	1220	470	75	24	270	2 Tan	Conv.
	15AC Sedan	53	Continental	145	2050	900	99	46	391	4	Conv.
(Champion)	7EC Champ	57	Continental	90	1450	630	87	35	357	2 Tan	Conv.
(Champion)	7ECA Citabria	70	Lycoming	108	1650	670	97	43	391	2 Tan	Conv.
(Champion)	8KCAB Decathlon	70	Lycoming	150	1800	570	109	46	491	2 Tan	Conv.
(Bellanca)	8GCBC Scout	75	Lycoming	180	2150	835	109	45	500	2 Tan	Conv.
American	AA-1 Yankee	70	Continental	108	1500	550	117	50	452	2 SBS	Tri
(Grumman)	AA-1B	75	Lycoming	108	1560	580	108	52	426	2 SBS	Tri
(Grumman)	TR-2	75	Lycoming	108	1560	525	110	52	443	2 SBS	Tri
(Grumman)	AA-5A Traveler	75	Lycoming	150	2200	950	128	50	591	4	Tri
(Grumman)	AA-5B Tiger	75	Lycoming	180	2400	1115	139	53	617	4	Tri
(American General)	AG-5B Tiger	93	Lycoming	180	2400	990	143	53	548	4	Tri
Aviat	Husky	93	Lycoming	180	1800	600	122	37	550	2 Tan	Conv.
Beech	Bonanza B35	50	Continental	165	2650	1060	148	49	652	4	Tri-Retract
	Bonanza K35	60	Continental	250	2950	1120	174	52	791	4	Tri-Retract

Make	Model		Engine								Gear
	Bonanza V35B	70	Continental	285	3400	1430	177	55	521/965	4–6	Tri-Retract
	Bonanza V35B	80	Continental	285	3400	1280	172	51	717	4–6	Tri-Retract
	Bonanza F33A	90	Continental	285	3400	1170	172	51	717	4–5	Tri-Retract
	A36	80	Continental	285	3600	1410	168	52	696	4–6	Tri-Retract
	A36	90	Continental	300	3650	1400	176	59	870	4–6	Tri-Retract
	Musketeer	65	Continental	165	2350	925	119	51	678	4	Tri
	Musketeer Sport	70	Lycoming	150	2250	900	114	49	661	4	Tri
	Sundowner 180	80	Lycoming	180	2450	950	104	51	600	4	Tri
	Sierra	80	Lycoming	200	2750	1040	137	60	652	4	Tri-Retract
	Skipper	80	Lycoming	115	1680	580	96	47	330	2 SBS	Tri
Bellanca	Cruiseair	50	Franklin	150	2150	900	143	38	591	4	Conv.-Retract
	Cruisemaster	50	Lycoming	190	2600	1070	157	36	591/869	4	Conv.-Retract
	260C	70	Lycoming	260	3000	1150	162	54	870	4	Tri
	Viking 300	70	Lycoming	300	3200	1250	169	61	717	4	Tri-Retract
Cessna	120	49	Continental	85	1450	660	87	36	391	2 SBS	Conv.
	140	50	Continental	85	1500	600	91	38	391	2 SBS	Conv.
	150	59	Continental	100	1500	540	105	42	452	2 SBS	Tri
	150	65	Continental	100	1600	630	106	43	426	2 SBS	Tri
	150	70	Continental	100	1600	625	102	42	391/630	2 SBS	Tri

(Continued)

Table 5-1. Used aircraft specifications (*Continued*)

Manufacturer	Model	Year	Engine Make	HP	Gross	Useful Load (lbs)	Cruise (kts)	Stall (kts)	Range (nm)	Seats/Config.	Gear
Cessna (*cont.*)	152	80	Lycoming	110	1670	530	107	43	321/547	2 SBS	Tri
	170	50	Continental	145	2200	1020	104	45	417	4	Conv.
	172 Skyhawk	57	Continental	145	2200	940	108	45	478	4	Tri
	172 Skyhawk	60	Continental	145	2200	940	108	45	539	4	Tri
	172 Skyhawk	65	Continental	145	2200	940	104	45	539	4	Tri
	172 Skyhawk	70	Lycoming	150	2300	1050	114	43	539/556	4	Tri
	172 Skyhawk	80	Lycoming	160	2400	980	120	46	435	4	Tri
	175 Skylark	60	Continental	175	2350	1040	121	45	517	4	Tri
	177 Cardinal	70	Lycoming	180	2500	1140	120	50	565	4	Tri
	Cutlass RG	80	Lycoming	180	2650	1100	140	50	722	4	Tri-Retract
	182 Skylane	60	Continental	230	2650	1120	137	50	574	4	Tri
	182 Skylane	70	Continental	230	2800	1160	139	50	600/791	4-6	Tri
	182 Skylane	80	Continental	230	2950	1310	143	48	783	4-6	Tri
	182 Skylane RG	80	Lycoming	235	3110	1300	157	50	843	4-6	Tri-Retract
	190	50	Continental	240	3350	1330	139	52	652	4	Conv.
	195	50	Jacobs	300	3350	1320	143	52	652	4	Conv.
	207 Skywagon	70	Continental	300	3800	1920	137	58	508/673	7	Tri
	207 Stationair	80	Continental	300	3800	1700	143	58	347/521	8	Tri
	210 Centurion	60	Continental	260	2900	1160	165	51	657	4	Tri-Retract
	210 Centurion	70	Continental	285	3800	1720	163	57	665/926	6	Tri-Retract
	210 Centurion	80	Continental	300	3800	1600	171	57	726	6	Tri-Retract

Manufacturer	Model		Engine								Gear
Ercoupe	Ercoupe Model E	50	Continental	85	1400	560	96	43	374	2 SBS	Tri
(Fornair)	Ercoupe	60	Continental	90	1400	510	105	43	435	2 SBS	Tri
(Alon)	Ercoupe	66	Continental	90	1450	520	108	43	391	2 SBS	Tri
(Mooney)	Cadet	70	Continental	90	1450	500	103	40	452	2 SBS	
Interstate	Artic Tern	80	Lycoming	150	1900	830	102	30	478	2 SBS	Conv.
(Arctic)	Arctic Tern	90	Lycoming	150	1900	830	102	30	478	2 SBS	Conv.
Lake	LA4	70	Lycoming	180	2400	800	114	44	543	4	Tri-Retract
	LA4-200	90	Lycoming	200	2690	1030	130	39	565	4	Tri-Retract
	Renegade	90	Lycoming	250	3050	1200	132	49	896	6	Tri-Retract
Luscombe	8F Silvaire	49	Continental	90	1400	590	100	42	391	2 SBS	Conv.
Maule	M4 Jetasen	70	Continental	145	2100	1000	130	35	609	4	Conv.
	M4 Rocket	70	Continental	210	2300	1080	143	35	591	4	Conv.
	M5 180C	80	Lycoming	180	2300	975	110	49	426/652	4	Conv.
	M-6 Super Rocket	90	Lycoming	235	2500	1000	139	30	426/747	4	Conv.
Meyers	125C	50	Continental	125	1675	580	123	0	435	2 SBS	Conv.
(Aero Commander)	200	60	Continental	240	3000	1130	170	48	608/1130	4	Tri-Retract
	200	66	Continental	285	3000	1060	190	47	1070	4	Tri-Retract
Mooney	Mark 20	57	Lycoming	150	2450	1035	143	50	652	4	Tri-Retract
	Mark 20A	60	Lycoming	180	2450	970	157	50	547/782	4	Tri-Retract
	M20C Ranger	70	Lycoming	180	2575	1050	153	50	870	4	Tri-Retract
	M20E Chaparral	70	Lycoming	200	2525	940	151	53	852	4	Tri-Retract
	M20F Executive	70	Lycoming	200	2750	1110	161	54	1000	4	Tri-Retract
	201	80	Lycoming	200	2740	1100	170	56	974	4	Tri-Retract
	201SE	90	Lycoming	200	2750	970	168	53	952	4	Tri-Retract

(Continued)

Table 5-1. Used aircraft specifications (Continued)

Manufacturer	Model	Year	Engine Make	HP	Gross	Useful Load (lbs)	Cruise (kts)	Stall (kts)	Range (nm)	Seats/Config.	Gear
Morrisey	2150	60	Lycoming	150	1817	690	117	45	457	2 Tan	Conv.
Piper	PA-18 Super Cub 95	50	Continental	95	1500	710	87	33	304	2 Tan	Conv.
	PA-18 Super Cub 105	50	Lycoming	108	1500	675	91	33	304	2 Tan	Conv.
	PA-18 Super Cub 150	60	Lycoming	150	1750	820	100	33	400	2 Tan	Conv.
	PA-18 Super Cub 150	70	Lycoming	150	1750	820	100	37	400	2 Tan	Conv.
	PA-20 Pacer	50	Lycoming	125	1800	820	109	42	465/569	4	Conv.
	PA-22 Tri-Pacer	60	Lycoming	160	2000	890	117	43	465/569	4	Tri
	PA-24 Comanche	60	Lycoming	180	2550	1075	139	50	696	4	Tri-Retract
	PA-24 Comanche	60	Lycoming	250	2800	1200	157	56	678	4	Tri-Retract
	PA-24 Commanche C	70	Lycoming	260	3200	1430	161	57	639/982	4	Tri-Retract
	PA-28-140C Cherokee 140	70	Lycoming	150	2150	940	116	48	456/630	2–4	Tri
	PA-28-180 Cherokee 180E	70	Lycoming	180	2400	1100	124	50	630	4	Tri
	PA-28-161 Warrior	80	Lycoming	160	2325	985	127	50	522	4	Tri
	PA-28-161 Warrior II	90	Lycoming	160	2440	1000	126	50	522	4	Tri

Manufacturer	Model		Engine							
	PA-28-181 Archer	80	Lycoming	180	2550	1130	129	53	487	4	Tri
	PA-28-181 Archer II	90	Lycoming	180	2550	1040	129	53	522	4	Tri
	PA-28-200 Cherokee Arrow	70	Lycoming	180	2500	1080	141	56	748	4	Tri-Retract
	PA-28RT-201 Arrow IV	80	Lycoming	200	2750	1110	143	55	722	4	Tri-Retract
	PA-28-236 Dakota	80	Lycoming	235	3000	1400	144	56	652	4	Tri
	PA-32 Cherokee Six	70	Lycoming	300	3400	1685	146	55	456/765	6	Tri
	PA-32-301 Saratoga	80	Lycoming	300	3615	1695	150	60	748	6	Tri
	PA-38-112 Tomahawk	80	Lycoming	112	1670	570	108	49	452	2 SBS	Tri
Rockwell	Lark Commander	70	Lycoming	180	2475	940	115	51	487	4	Tri
	Commander 112A	75	Lycoming	200	2650	960	140	54	848	4	Tri-Retract
(Commander)	112B	93	Lycoming	260	3250	1200	164	57	674	4	Tri-Retract
Ryan	Navion	50	Continental	185	2750	970	135	47	434/652	4	Tri-Retract
Stinson	180 Voyager	49	Franklin	165	2400	1110	113	44	478	4	Conv.
Taylorcraft	F-19	50	Continental	65	1200	450	83	33	304	2 SBS	Conv.
	F-22	90	Lycoming	118	1750	660	102	36	557	2 SBS	Conv.

Note: These figures are as released by the indicated manufacturers at the time the aircraft were built. The different companies often published different sorts of figures. Some published a range of dry tanks; others included a 30-minute VFR or 45-minute IFR reserve.

Now, this is by no means a complete list. It's just a sample of some of the airplanes available. I've tried to pick out planes that had longer production runs, or evolved into other airplanes. Undoubtedly, I've left off someone's favorite. Sorry about that.

There is sometimes a considerable amount of optimism evident in the performance levels in Table 5-1. I'm afraid the marketing departments were just as responsible for these numbers as the flight test section. I really chuckled when I learned that Cessna claimed that my old 150 cruised at almost 120 mph. Then again, 30-year-old airplanes are not going to perform to as-new specifications. In other words, take the table as a guideline, not gospel.

Here's an explanation for the columns on the table:

Manufacturer. The name of the company that *first built* the model of aircraft. Later companies making the same plane are listed in parentheses under the original manufacturer. For instance, look under "Ercoupe." You'll see the specifications for the original Ercoupe Model E, as well as the Ercoupes built by Forney and the modified version sold by Mooney.

Model. The model name/number the aircraft was sold under, as well as its "street" name. For instance, to a Cessna purist, the Cessna 172 and the Cessna Skyhawk are two different Cessna models—the Skyhawk was the upgraded version, with upgraded avionics and standard equipment. The table usually gives the performance values for the fancy version.

Year. The year of aircraft, for which the performance figures are given.

Engine Make. The company which manufactured the engine.

HP. The engine's rated horsepower.

Gross. The allowable gross weight.

Useful Load. The difference between the gross weight and the manufacturer's listed empty weight. It is a dead certainty that any aircraft you look at will have a lower useful load than that listed here.

Cruise. The claimed cruise speed (at 75 percent power) in knots.

Stall. The stall speed in statute miles per hour.

Range. The published range, in nautical miles, when flown at 75 percent power. Where two values are given, the second one indicates the range when optional larger fuel tanks are installed.

Seats/Config. The number of passengers the plane can legally carry. If the aircraft is a two-seater, SBS stands for a side-by-side seating configuration, and "Tan" means tandem.

Gear. The type of landing gear. "Conv." means conventional (Taildragger), and "Tri" means tricycle. "– Retract" means the aircraft has retractable gear.

Now that we have some background on aircraft data, let's take a look at the used-airplane buying process.

DETERMINING THE DOLLARS

Between the performance, capability, and features selected in Chap. 4, and the table of used aircraft, you've got a pretty good idea about what type of airplane to buy. The next step is to determine, which models fall within your price range.

Finding the prices

With used cars, determining typical selling costs is relatively easy. One can scan the classified ads and keep a running mental tally of the price range. Or, go to the bank and thumb through their "blue book" of automobile prices. You can, to a significant extent, do the same things with airplanes. However, there are a few differences that get in the way.

First, your local newspaper is not likely to have a wide variety of aircraft for sale. If you wanted to determine what '85 Pontiacs sell for, you're in luck. But there are just not enough aircraft-for-sale listings in your typical newspaper to determine a price range for a given model. To make matters worse, the major publicly-available aircraft price guide is apparently no longer published.

These days, the Internet is a good start. Just enter "aircraft for sale" on your favorite search site (Google, Yahoo, and the like.) and see the screen flood with listings. Or, check with EBay or other online auction sites. There are also a number of aviation publications with extensive "For Sale" sections. The granddaddy of them all is Trade-A-Plane. This tabloid is published three times a month, and includes aircraft ads for all around North America. Many consider it the standard for determining aircraft prices. It's available at most aviation supply stores. For a subscription, contact:

Trade-A-Plane

P.O. Box 509

410 West 4th St.

Crossville, TN 38557

Subscribers can also access the listings online at *http://www.trade-a-plane.com*.

An Internet search with "aircraft valuation service" as the search term will present several sources, most requiring a fee. However, Aircraft Owners and Pilots Association (AOPA) does offer this service free to members (*www.aopa.org*).

The want-ad trap

There is a rather insidious hazard involved with reading aircraft-sales ads. Say you're interested in a Cessna 172 in about the $30,000 range. You'll read through the ads, and find a couple of good prospects. But then you'll stray to another category. Hmmm . . . here's a '63 Mooney for only $25,000. Or gee . . . a Cessna 310 twin for only a couple of thousand more. Not so very long ago, I found a DC-3 in flyable condition (supposedly) for $38,000!

As we discussed in Chap. 3, the operating cost of a used airplane is not based on its purchase price. The operating cost is determined by how much the plane would sell for *if it were new*. However, in the real world, it's not a bad idea to have an alternate aircraft in mind. If you're looking for a Cessna 182, for instance, remember that a Piper Archer's performance and features are somewhat similar. Or maybe something off the beaten path might fit your needs (Fig. 5-4). Do a little pricing on an alternative or two, just to be able to tell if a good deal arises unexpectedly.

Why? Maybe, just maybe, you'll latch onto a bargain. Unless you've got the hots for a particular make and model, the ability to change targets might prove useful. By doing your homework on a couple of alternatives in advance, you might be able to get more bang for the buck.

Financing

The primary factor at this point is how much money you can afford to spend. If ready cash is burning a hole in your pocket, this part is relatively easy. However, if you plan on financing, it's best to lay the groundwork prior to starting your search. A seller may not want to wait while your application works its way through a financial institution.

I had a prime example of this when I sold my 150. One younger man came by for a test flight, and really liked the plane. He was a first-time buyer, enthusiastic about my little red-and-white bird. But he didn't have the money available, yet. Would I hold the plane for him? I declined, since I'd figured it'd sell relatively fast. He said he'd give me a call when he got the money together. Sure enough, he called a week after the plane sold. He was crushed. And I'll be honest, it hurt me as much as it did him. But then, he called six weeks after our original conversation. I don't know why it took him that long to raise the money. Maybe he had to sell a car, first. But if he'd handled the money details *before* he started looking, he'd have made us both happy.

Fig. 5-4. *Stinson 108-3 Flying Station Wagons have four seats and an excellent useful load—they can make an excellent alternative to more-modern aircraft like the Cessna 182.*

As far as financing, you have three choices: A loan with the aircraft as collateral, one with another item of property securing the loan, or an unsecured loan (sometimes called a "signature" loan).

You can finance an airplane just like a car, with the vehicle itself acting as collateral. Maximum loan periods will range from 5 to 15 years, depending upon the age of the aircraft and the amount financed. Seven seemed the typical maximum for most small-aircraft loans. You can go shorter, of course.

Expect to put at least 10 percent down on the aircraft. Finance rates seem to run about the same as car loans. Contact the company in advance to lay the groundwork, and most can close a loan within 48 hours of your finding your dream plane. There are several problems potential borrowers may encounter. Some companies limit the age of the aircraft they'll finance. With the average lightplane being over 30 years old, it's obvious that finding loans is going to be tough for almost half the General Aviation fleet.

Some institutions won't make aircraft loans for less than $25,000. This leaves out a number of the typical first-timer airplanes. The reason for this limitation is interesting. Loans of less than $25,000 are considered to be *consumer* loans. Consumer-protection laws kick in. Such loans have a much-higher paperwork involvement; as one financial officer told me candidly, "We have to treat the customer as an unsophisticated borrower."

Ouch!

To find a company that finances aircraft, just enter "aircraft loan" on your favorite Internet search site. AOPA also has an aircraft-finance program. Local banks might be willing to handle your purchase. A Bonanza-owning friend of mine took out a $35,000 loan with a neighborhood lender, and later added it to finance his Bonanza's engine rebuild. My discussions with several local banks and savings and loans indicated aircraft loans weren't unknown to them. One bank manager was quite positive about lending money for airplanes. My credit union offers airplane loans at 8.25 percent interest for up to 15 years.

Of course, one needn't take out a loan on the aircraft itself. A lot of planes are financed through loans secured in other ways, such as a home equity loan. These types of loans have one big advantage over the traditional aircraft loan—aircraft loans require a 10–20 percent down payment. When the aircraft doesn't act as collateral, you can borrow the full amount necessary, and keep your cash for a maintenance reserve. Or even borrow $10,000 or so over the cost of the plane, to cover immediate expenses.

GET SET

Plane's picked out. Money's settled. Just a couple more things to take care of before leaping into the used-airplane market.

Knowing the warts

No designer is perfect. All airplanes have, over time, revealed certain problems. And problems equal money, as far as the aircraft owner is concerned. As mentioned in Chap. 3, *Aviation Consumer* magazine performs in-depth reviews of various small aircraft. These reviews include a good summary of the aircraft's maintenance tendencies. You'd like to know what these tendencies are. So when you're looking at a prospective purchase, you can concentrate on these trouble areas. The magazine's reviews have been collected and published in *The Aviation Consumer Guide to Used Aircraft*.

Serious aircraft mechanical faults are the subject of FAA Airworthiness Directives. Compliance with ADs is mandatory, yet a few owners seem to sneak by. You surely don't want to buy their plane and end up paying for the missed ADs. A little research is in order. The best way to search for ADs is online. Go to www.faa.gov and select "Airworthiness Directives" on the "Quick Find" pulldown menu. If there is an online type club for the type of plane you're contemplating buying, check it out—they may well have an AD section. You can also research the applicable ADs at an FAA Flight Standards District Office(FSDO) or the base of an agreeable A&P with Inspection Authorization (IA). IAs are required to maintain an AD file, but they aren't obligated to let the general public paw through their listings. I talked to a couple of FBOs who said they wouldn't be too adverse to giving access to their AD list. If your local IA is agreeable, don't abuse their good nature. And give them some of your maintenance business after you buy.

When searching, look under the manufacturer's name, then find the model of aircraft you're interested in. There'll be two listings of interest. The first is the *Series* heading (or model/series on the web page). This leads a set of Ads, which apply to the entire series of aircraft, no matter the year or model number. After the series heading, the individual models and years of the aircraft type are listed. Both sets of ADs apply to your prospective purchase (Fig. 5-5). But the listings under the aircraft type include only those Ads, which apply specifically to the airframe. There are separate sections for the engines, propellers, and appliances.

You may not know what engine and propeller to look at. Sure, you'll know a mid-70s 172 has a Lycoming O-320, but both the O-320-D2J and O-320-H2AD were used. You can find out exactly, which engine and propeller a particular aircraft used by consulting its Type Certificate Data Sheet. A sample first page of a data sheet is shown in Fig. 5-6. The FSDO and your friendly IA will have copies of these, as well.

Getting back to ADs, each is identified by a group of three numbers, such as 82-10-03 or 64-20-01. The first two digits are the year in which the AD was published. The second indicates in which "series" the AD was published; these series are generally released biweekly. Finally, the last two digits indicate where this AD was included in the series.

With a list of the ADs applicable to your prospective purchase, what's your next move?

Well, you could look up the ADs to find out what they cover. There may be a subject or two that piques your interest. If the subject was "Wing Attach Bolt Failures," you'd undoubtedly want to find out more. And jolly quick!

One item of interest is whether the AD is "one-time" or "recurring." An owner's involvement with one-time ADs ends when the steps described in the notice are performed and logged. A typical example is the replacement of a flawed component. A recurring AD must be performed at intervals based on aircraft or calendar time: "Inspect the bellcrank using a dye penetrant every 200 hours," for example.

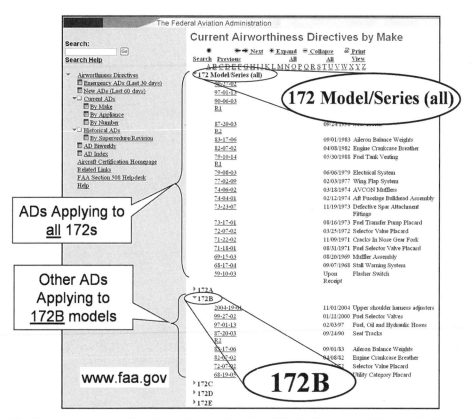

Fig. 5-5. *An example of the Airworthiness Directives (ADs) listing on the FAA web site. If one is looking for the ADs for a Cessna 172B, both the ADs linked to the "172 Model/Series" link and the "172B" link apply.*

You're probably going to be more concerned with the recurring ones than the one-timers. You'll check that the one-time ADs have been complied with, of course, but if they've been performed they won't cost you anything more. On the other hand, looking up the AD text at this point might just be a wasted effort. After all, the mechanic performing the prepurchase inspection on your prospective aircraft will be checking the same paperwork. All you really need is a listing of what ADs *should* have been done, so you can weed out poorly-maintained candidates. The index supplies the AD number.

Then again, if you're there at the bookshelf (or microfiche projector, as shown in Fig. 6-6) and have the time, indulge yourself. There's a lot of interesting stuff in the AD listings. If you haven't decided on a particular model yet, the AD listing might aid the decision. A sample AD is included as Fig. 5-7.

Service bulletins and Service Difficulty Reports (SDRs) are also of interest. Service bulletins are published by the aircraft manufacturer. They're similar to, but complying with them is not required by regulation, though the manufacturer may term them "mandatory." Often they're the precursor to an AD. These can usually be found online at the manufacturer's web page.

DEPARTMENT OF TRANSPORTATION
FEDERAL AVIATION ADMINISTRATION

A13CE
Revision 24
Cessna
177
177A
177B
March 31, 2003

TYPE CERTIFICATE DATA SHEET NO. A13CE

This data sheet which is part of Type Certificate No. A13CE prescribes conditions and limitations under which the product for which the type certificate was issued meets the airworthiness requirements of the Federal Aviation Regulations.

Type Certificate Holder	Cessna Aircraft Company P O Box 7704 Wichita KS 67277

I. Model 177, Cardinal, 4 PCLM (Normal Category), approved February 16, 1967
2 PCLM (Utility Category), approved August 8, 1967

Engine	Lycoming O-320-E2D
*Fuel	80/87 minimum grade aviation gasoline
*Engine limits	For all operations, 2700 rpm (150 hp)
Propeller and propeller limits	McCauley 1C172/TM Diameter: not over 76 in., not under 74 in. Static rpm at maximum permissible throttle setting: not over 2360, not under 2260 No additional tolerance permitted.
*Airspeed limits (CAS)	Never exceed 185 mph (160 knots) Maximum structural cruising 145 mph (125 knots) Maneuvering 113 mph (98 knots) Flaps extended 105 mph (91 knots)
C.G. range	Normal category: (+101.0) to (+114.5) at 2000 lbs. or less (+105.5) to (+114.5) at 2350 lbs. Straight line variation between points given. Utility category: (+101.0) to (+109.9) at 2000 lbs. or less (+103.6) to (+109.0) at 2200 lbs.
Empty weight C.G. range	None
*Maximum weight	Normal category: 2350 lbs. Utility category: 2200 lbs.
Number of seats	4 (2 at sta. +93.0, 2 at sta. +134.0)
Maximum baggage	120 lbs. (+162.0)
Fuel capacity	49 gal. (two 24.5 gal. fuel bays in wing at sta. +112, 48 gal. usable) *See Note 1 for data on system fuel.*

Page No.	1	2	3	4	5	6	7	8	9	10	11
Rev. No.	24	22	21	23	20	20	20	21	21	23	24

Fig. 5-6. *The Type Certificate Data Sheet defines the exact configuration of each model of aircraft.*

Airworthiness Directive

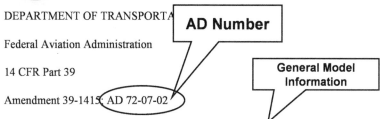

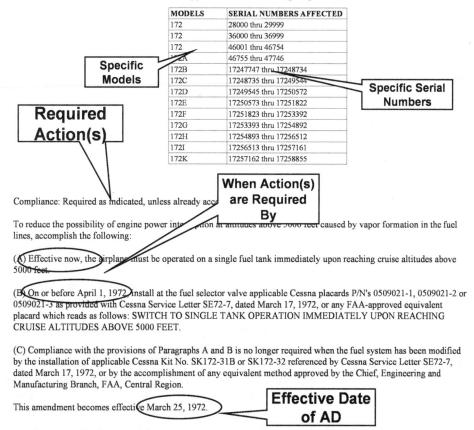

Fig. 5-7. *A sample AD note. It specifies the applicable aircraft model and serial numbers, the compliance period, and the work to be performed.*

Whenever A&Ps discover unusual problems during maintenance, the FAA encourages them to file SDRs. These are compiled and published periodically. A typical SDR may call attention to an abnormal wear pattern or a peculiar anomaly in a particular aircraft. Often they include a recommended method for preventing or solving an occurrence on other aircraft. SDRs aren't mandatory either; however, the FAA tracks them to try detect patterns. Often an AD results if a certain problem occurs in a number of aircraft.

An Internet search for "FAA Service Difficulty Reports" should do the trick. Obviously, researching the AD, service bulletins, and SDRs can be time-consuming. It's kind of fun, though—sort of like playing aviation detective.

Knowing the words

When you learned how to fly, you had to learn a new language. Say "Ground Point Niner off the active" to someone at the bus stop. You'll end up with a seat all to yourself. Works great.

It's no different when buying aircraft. There's a whole new terminology to absorb; one based on abbreviations and shorthand.

Table 5-2 defines a set of common aircraft-sales terms and abbreviations. Some require a bit more explanation:

DAMAGE HISTORY (NDH, NMDH, MDH, NRDH)

This indicates whether the aircraft has ever required airframe repair for any reason—from a flight accident to a hangar door falling on it.

No Damage History (NDH) is easily understood. No Major Damage History (NMDH) indicates that some slight repairs had been necessary at some point. For a metal airplane, this usually indicates that some metal had to be repaired or replaced.

Major Damage History (MDH) means the plane got seriously bunged up at some point. Listing a plane as MDH in an ad takes a certain amount of guts; there's obviously something in the logs that no amount of salesmanship can hide. 14CFR Part 43 defines the operations, which count as "major repairs." Unfortunately, some mechanics and sellers have their own definitions of what constitutes "major" damage.

No Recent Damage History (NRDH) indicates the plane has some old repairs. It's not quite as bad as MDH, as it indicates the repairs were successful enough to allow the plane to continue flying for a considerable time. Still, it hurts the value of the airplane. Given your druthers, you'd prefer a plane that's never felt the mechanic's rivet gun or welding torch. Given the age of the General Aviation fleet, though, you may have to settle for a well-repaired bird. That's the key, of course—well-repaired. Such repairs are one item your mechanic should examine during a prepurchase inspection.

Aircraft come under more scrutiny during the annual if they've ever been subjected to a "major repair." Check with your local A&P if you find a hot prospect listed as MDH.

Table 5-2. Aircraft advertising terms and abbreviations

ADF	Automatic direction finder
AFTT	Airframe total time
AP or A/P	Autopilot
Auto Gas STC	Approved to run on car gas
CH	Channel (i.e., "760 CH Comm radio)
CS or C/S	Constant speed propeller
DME	Distance measuring equipment
EGT	Exhaust gas temperature instrumentation
Enc	Transponder altitude encoder
FD	Flight director
GPS	Global postioning system (satellite navigation) receiver
GS	Glide slope
HC	Heavy case*
HSI	Horizontal situation indicator
MB	Marker beacon
MDH	Major damage history*
N/C	Nav-comm radio
NDH	No damage history*
NMDH	No major damage history*
NRDH	No recent damage history*
RNav	Area navigation equipment (generally obsoleted by LORAN and GPS
720	720-channel communications radio
SCMOH	Since chrome major overhaul*
SFRM	Since factory remanufactured engine*
SMOH	Since major overhaul*
SPOH	Since propeller overhaul
STC	Supplemental type certificate
Stits	Covered in stits-brand fabric
STOH	Since top overhaul*
TBO	Time between overhauls
STOL	Short takeoff and landing modifications
TT	Total (flight) time
TTAE	Total time airframe and engine
TTAF	Total time on airframe only
TTSN	Total time since new
XPDR, Xponder	Transponder

*Further explanation in text.

TIME SINCE OVERHAUL (SMOH, SFRM, STOH, SCMOH, SCTOH)

The biggest cost driver on any small aircraft is the condition of the engine. Any ad should make some reference to the number of hours the engine has run since it was new or since its last overhaul.

The manufacturers of certified engines assign a recommended Time Between Overhauls (TBO) for their products. This TBO only legally applies to aircraft in commercial service. An owner may fly his or her aircraft beyond its engine's TBO. Even so, it's a risk. The fact that the engine has a lot of running time since a detailed internal inspection hurts its value. And hence the value of the plane.

Two types of engine times are tracked—the Total Time (TT) and the time Since Major Overhaul (SMOH). An engine listed as "3500 TT, 850 SMOH" has a total of 3,500 flight hours and was overhauled 850 hours ago. You won't see the TT in two cases. The first is if the engine has never been overhauled. "1100 TTSN" might be a typical listing. The second is if the original engine maker has remanufactured the engine. An ordinary overhaul requires that parts worn past a certain point, called the *service limits*, be replaced. When the maker remanufactures the engine, only parts which meet or beat new-engine tolerances are kept. The original engine log is thrown away, as is any time the engine had accrued up to that point. The engine on a used aircraft might then be listed as "450 SFRM," or 450 hours Since Factory Remanufactured. These engines are also referred to as "zero-timed."

Independent shops can also rebuild engines to new limits, just like remanufacturing. However, the FAA only allows the original engine manufacturer to "zero-time" engines. An independent company's "new limit" rebuild might be of higher quality than the factory's remanufacture. But the independent must still retain the engine's accumulated time.

Cylinders and valves seem to cause the most problems on aircraft engines. Often, they require replacement or a rebuild prior to the rest of the engine. A "top overhaul" restores the cylinders, pistons, and valve train to at least service limits. Prospective buyers might then see an engine listed as "2600 TT, 1200 SMOH, 10 STOH," 2600 hours since the engine was new, 1200 hours since its last major overhaul, and 10 since the top overhaul.

Chrome-plated cylinders are sometimes used in the overhaul process. The chrome plating increases the hardness of the surfaces, reducing wear significantly. However, the process is expensive and sometimes difficult to break in. The "C" in listings, such as 400 SCMOH or 120 SCTOH, means that the cylinders were chromed at the last overhaul/top overhaul.

HEAVY CASE

One of the disasters that can happen to a small aircraft engine is a crack in the crankcase. At worst, the crankcase must be scrapped. At best, the engine must be totally disassembled so the crack can be welded by a specialist.

If a certain engine seems prone to crankcase cracking, the manufacturer will often design a case with thicker walls. Such engines are preferable. Your research should indicate, which models of your target aircraft used the thinner-case engine. However, owners, when faced with a case repair, sometimes replace the entire engine with the improved model. When the airplanes go on sale, the owners will indicate the presence of the heavier-cased engine.

The thin-walled versions aren't the kiss of death, though. One school of thought says that if the engine makes it through a full run to TBO without cracking, it probably never will. The cracking is sometimes initiated by unrelieved manufacturing stresses, and an engine that goes 1500 hours or more without cracking probably doesn't have the stress points.

Other preparations

Get some insurance quotes based on your target aircraft. Work out with the agent or representative about how to initiate a new policy. You're going to be flying your new toy a *lot* right after the purchase. Make sure you'll be able to get an immediate start on your insurance coverage.

If you want a hangar and haven't got your name on the list, do so now. Why did you put it off so long? A friend of mine spent five years on a waiting list before a hangar came open at the closest field. In the meantime, you can probably find outside tiedowns available in most areas. There's never been an airplane made that couldn't stand a month or two in an open tiedown. Figure out in advance where you're going to keep your new purchase, for the first couple of weeks at least.

As mentioned earlier, you may have to pay a local tax or fee on an aircraft purchase. You might do some scouting out in advance on how to minimize this. Just taking possession in a different county may make a difference.

Finally, download a copy or two of Form AC 8050-2. This is the aircraft Bill of Sale. There'll be a lot more forms to fill out afterward (surprise, surprise), but the Bill of Sale is the only one that must be signed by the seller. Might as well download a copy of Form AC 8050-1 (Registration application) while you're at it.

As of this point, you're ready to start actively searching.

THE HUNT

Where are you going to find your dream plane?

In a perfect world, it'll be sitting on the ramp at the airport you normally fly out of. You'll be walking back from another session in a clapped-out rental airplane, and there it'll be. A faded "For Sale" sign maintains a yellowed scotch-tape cling inside the window. The model is exactly the one you're looking for. The price is right in your range. And you recall seeing it running around the pattern the previous weekend (Fig. 5-8).

Why was this situation so ideal?

1. You can take all the time needed to examine the airplane. If it's getting dark, you can come back the next day. If the plane were at an airport 200 miles away, you'd feel pressured to decide yes/no immediately to avoid another long trip.

2. Since it's your home airport, you'll probably have a good idea where to find a mechanic to perform a prepurchase inspection.

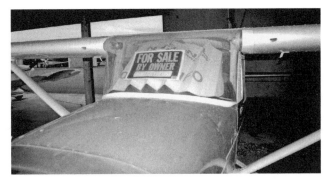

Fig. 5-8. *Sometimes the most fun way to shop for an airplane is to wander around airports looking for "For Sale" signs.*

3. The faded sign indicates that it's been on the market for a while. It could be overpriced for its condition, or perhaps the owner doesn't want to pay for any advertising.

4. If you get to the point where you're ready to test fly, you'll be operating in familiar airspace. You'll be able to concentrate on checking out the plane rather than navigating over unfamiliar ground.

5. The fact that you've seen it flying is a sign that the aircraft is (probably) airworthy.

Such an ideal situation probably won't happen. Still, there are some lessons, here.

Hints for the hunter

Start locally. Checking out a prospective purchase is far easier when the plane is based nearby. Wander the local airports. Many now have security fences to keep out the riff-raff, but your pilot's license should be sufficient for entry.

Check the bulletin boards (Fig. 5-9). There's usually one at the airport office, at any FBO on-field, and outside the airport restaurant. Note that such a search is a good justification for renting an aircraft on a nice Saturday afternoon. Instead of the famous $50 hamburger, you could be searching for the $40,000 Piper. Use the phone, too. Local libraries carry the yellow pages for nearby cities. Look up the FBOs and give them a call.

Other options

There's nothing wrong with consulting the online advertisements, or checking the print ads like those in "Trade-A-Plane" or "General Aviation News," especially if you're looking for a very specific aircraft or level of equipment. Unfortunately, a lot of the aircraft will be out of convenient range. If you live in Texas and find a plane for sale in New Hampshire, what next? Unless you have a flying friend up there, you'll have to travel to New England to examine the aircraft. One wasted trip

Fig. 5-9. *When you're in the market for a plane, airport bulletin boards are good excuses to visit nearby fields.*

just drives up the effective price you pay for whatever plane you eventually buy. Do an Internet search for "Area Codes" to find the geographic location of a particular code.

Buying a plane through a broker is a possibility. It's their business to know where available planes are. Plus, they should be able to help you find a loan, if necessary. Most of us don't buy too many planes in a lifetime; it can be useful to get the assistance of someone who buys an airplane every week. However, remember that a broker's income stems from people who buy airplanes. Not folks holding out for exactly the right machine. Nor those who take a leisurely approach to the hunt. Like any field, aircraft brokerages have their share of unethical people. Some can be less than truthful about the equipment or condition of the planes they're trying to sell. Get everything in writing, and arrange your own prepurchase inspection.

Finally, while a broker can help you find the right airplane, the price will invariably be higher. And since their commissions are proportional to the value of the airplane, don't expect much help if you're in the Tripacer/170 sort of market.

Hangar queens and seized birds

There are a couple of special cases you might come across during your search. The first are Ramp or Hangar Queens. You see these aircraft at any airport—sagging, covered with dust, obviously not having flown for years (Fig. 5-10).

Why are they so neglected? Any number of reasons. The owner may have lost his or her medical, but still holds out the hope of getting it back. Perhaps money is tight, and they're just hanging onto it until times get better. Maybe there's something wrong with the plane and the owner just can't afford to fix it at present. You'll look at the plane, and a little thought will tickle the back of your brain. "Maybe I can pick it up for a song. A little TLC, and I can be flying."

Fig. 5-10. *Close-up of a "Ramp Queen." The tire isn't just flat; it had sank into the soft asphalt. Moss had grown on the underside of the wings. While you might be able to buy the aircraft itself for little money, getting it back into flying shape will be expensive.*

There are two sure-fire ways to hurt an aircraft. One is to perform a Split-Ess from 200 feet. The second is to park the plane and abandon it. Neglect does major, big-time damage to airplanes. Engines rust internally, roughening the cylinder walls and bearings. Moisture and left-over combustion products produce acids, which attack metal. It doesn't take much idleness to cause problems. The TBO on Lycoming engines, for instance, is based on 25 hours of operation per month. Consider that many, perhaps most, General Aviation aircraft operate a hundred hours a year or less.

It isn't just the engine, either. Moisture seeps into the cabin, rotting upholstery and carpets. It oozes into instruments, rusting them from the inside. Control bearings freeze. Pulleys lock. And our feathered and furred friends get to work. At my home field, I'd watched a Globe Swift sit out in the open for years. The owner stubbornly hung onto it. The tires went flat. Pieces disappeared. The windows crazed and opaqued. Finally, someone talked the owner into selling and began the restoration process. I was there when he opened up one of the inspection panels in the wing, reached inside, and starting pulling out handfuls of bird and mouse nests. And not one of them little critters had been "plane-trained." Avian and mammalian urine and excreta (the new owner had other names for it) will corrode anything it comes in contact with.

That's not to say you can't make a good deal on a hangar or ramp queen. But it'll need more than just a cursory inspection before you fly it. You might luck out. Or you might end up with a nine-year restoration, like a case study in a later chapter.

Another special case you'll run into is an aircraft without logbooks. Or maybe just the engine or just the airframe logs are missing. They might have been repossessed, or seized from a drug smuggler and up for bids at a government auction.

Missing logbooks is a sticky situation. You won't know the time on the airframe or engine. You'll have utterly no idea of the history of the plane. Worst of all, the logbooks are used to record compliance with ADs. At a minimum, you'll have to have an A&P verify or perform every AD that's *ever* been published on the airplane.

As you can imagine, that would be a king-size hassle. It might be worth it, if the purchase price was low enough. But talk to a mechanic in advance; get an idea of how much money and paperwork is going to be required. But, there is a happy medium. A decrepit, abandoned airplane can be more hassle than it's worth, but if the plane is still in somewhat decent shape, fixing it up can be less of a problem. This approach isn't for the faint of heart—or for those who don't have the disposable income to fund the upgrades. But it does have the advantage of letting you configure and equip the airplane exactly the way you want it.

A major factor in the value of an airplane is the condition of the engine. Most buyers steer away from planes with higher-time engines. However, replacing an engine is a relatively simple (albeit expensive) process. If the sale cost of the airplane is low enough because of the run-out engine, it can make economic sense to buy the airplane and exchange the engine for a rebuilt one. The big advantage of this approach is that you now have an engine that's a known quantity—you *know* the engine's condition, and won't have to worry that the previous owner had been hiding problems. It's a lot of effort, but it can work out nicely.

CASE STUDY: A SEVEN-LEAGUE CESSNA 150

When a prospective buyer's mission statement calls for long cross-countries, the lowly Cessna 150 usually is about the first plane to drop off the list of candidates. But Miles Erickson ignored conventional thinking. He'd been flying planes with a local flying club, but the kind of long trips he wanted to take just weren't feasible in a rental airplane. There was always the potential for the weather to close in, and costs skyrocket if one can't return a rental airplane on schedule.

"The best thing about the Cessna 150," says the Seattle urban/transit planner, "Is the low operating costs, making long-distance travel reasonably affordable." His first long trip was to bring the airplane home. He found N3913J, a 1967 150G (Fig. 5-11), via the Internet in Chico, California in spring of 2003. It had a basic set of avionics and new paint.

Erickson didn't have a prepurchase inspection performed. "It had just had an annual inspection," he says. "I felt comfortable with the previous owner, who had obviously taken very good care of his plane, and felt like I was clear on what the outstanding issues were." He paid $19,500 for it—a down payment came from his savings, and a family loan covered the rest, with a monthly payment of $500. He also had to pay another $1700 in use taxes when he brought the plane home to Washington state.

Ownership experience

Erickson puts about 200 hours per year on Cessna One Three Juliet. He's been all along the West coast in the airplane—business trips to Los Angeles and vacation

Fig. 5-11. *Small two seaters like Cessna 150s and 152s aren't usually thought of as airplanes for long-range travel, but Miles Erickson takes his on thousand-mile trips. (Courtesy Cessna Aircraft Company.)*

trips to Alaska. Since his destinations in Alaska often have gravel runways, he has added a rubber rock deflector (from the Cessna parts catalog) to the horizontal stabilizer to help protect the paint. He takes other long cross-countries as the opportunity arises, as well as local sightseeing flights, and trips to visit nearby friends.

The plane is kept at King County International (KCI) Airport (generally known as Boeing Field), just inside Seattle city limits. An outside tiedown costs Erickson $90 per month. Other local airports have lower tiedown rates, but KCI is just a 20-minute drive—or 30-minute bus trip—from Erickson's home. The ability to ride the bus to and from the airport lets him avoid leaving his car unattended during his long trips.

The 150 has an auto fuel Supplemental Type Certificate (STC), and Erickson estimates that he uses auto gas about a quarter of the time. "Sometimes I carry 5-gallon cans of mogas in at Boeing Field, but of course it's usually impractical to find auto gas at enroute stops on long cross-country flights." He uses a "Mr. Funnel" fuel filter to filter the fuel of both particulates and water, and sets the 5-gallon fuel can directly atop the wing. A self-starting safety siphon lets him avoid trying to "aim" the can.

A friend recommended his A&P, and Erickson participates in all maintenance, not just the annual inspections. The inspections themselves cost around $400. He's been hit by a number of minor (yet irritating) problems since buying the aircraft. Both the radio and the transponder have had to be replaced, and some other electrical problems have been traced to corrosion. Otherwise, the problems are typical for a 35-year-old airplane: flap track rollers, nose strut rebuild, tire replacements, a brake line, and the usual wear items.

The outside tiedown has lead to some paint deterioration. The airport is located in a semi-industrial area, but it is unknown whether air conditions are contributing

1967 Cessna 150G
Owner: Miles Erickson

Tiedown: $90/Mo
Insurance: $500/year
Hours/year: 200

Inspections: $400
Other yearly $3000*
maintenance:

*Includes replacing both the radio and transponder

Fig. 5-12. *A summary of Erickson's ownership costs.*

to the paint issues. His insurance costs are surprisingly low. Being a 500-hour private pilot, he pays only about $500 per year for liability and in-flight hull coverage.

Figure 5-12 presents a summary of Erickson's ownership costs.

Advice

Erickson bought the airplane without a prepurchase inspection, but he doesn't recommend it. "This worked out fine for me, but having heard some other folks' stories, I doubt that I'd ever buy a plane again without taking it apart and putting it back together with my own bare hands."

When he bought the airplane, one of the known problems was a cylinder with relatively low (but still legal) compression. After a year of flying, using autogas when possible, usually flying two and three-hour legs at high altitudes and/or economy power settings with aggressive leaning, the "bad" cylinder's compression problem disappeared. Erickson says, "Maybe that's another lesson learned—planes like to be flown!"

"Older avionics are virtually guaranteed to cause problems," he adds. After suffering a charging-system failure, he also cautions owners not to put their eggs in one basket: "Redundancy in the panel is no help in the case of an electrical failure . . . I strongly recommend carrying redundancy in your flight bag, instead. Not to mention lots of spare batteries!" He carries a handheld radio and a portable Global Positioning System (GPS) receiver on all his flights.

When you get down to it, a low-powered aircraft like the Cessna 150 is still faster than highway travel, especially for Erickson's flights to Alaska. A good used 150 can be bought for not much more than a new Toyota sedan. For Erickson, the small Cessna was a logical choice: "It's all the airplane I need, and I can afford it."

6

Special Wings, Part I: Antiques and Classics

The nonflying public is pretty funny. To them, all little planes are "Piper Cubs" and accidents happen because "the pilot didn't file a flight plan." They're strange in another way, too. Offer to take someone for their first airplane ride, and they'll likely say, "You'll never get *me* up in a little airplane!" Then some add: "But I've always wanted to take a ride on an old open-cockpit biplane."

Talk about mixed up! They won't risk their lives in an easy-to-fly tricycle-geared plane with a reliable modern engine. However, they dream about flying in a cranky 70-year-old taildragger powered by an engine built by a company that went bankrupt before they were born.

Still, though—one can't really blame them. Many pilots would love to fly a fabric-covered biplane from a fresh-mowed field. If some mad fool offered me the choice between an old Stearman or a brand new Bonanza, I'd be strapping on helmet and goggles before the word "or" left his lips.

Dreaming is one thing. What about reality?

That's what this chapter is about. We're going to take a look at some of the joys and difficulties of owning an antique or classic airplane.

First, let's establish what those names signify. Generally speaking (Fig. 6-1), an Antique is a plane built before World War II (WWII). A Classic is an aircraft built between the end of the war and up to about 1955.

A third category—Contemporary Classic—fills an interesting niche. This category covers planes built between 1955 and 1965, and for the most part, it covers the earliest models of planes like the Cessna 150, 172, 182, and Piper Cherokees.

Let's take a look at Antiques, first.

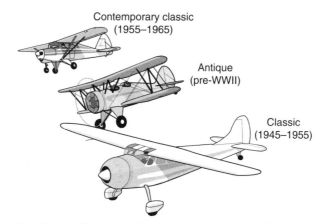

Fig. 6-1. *Antique and Classic aircraft are two different categories, with "Contemporary Classic" as an offshoot of Classic.*

ANTIQUES

What's it like to own an antique airplane? Probably the best analogy I can make is to old cars (Fig. 6-2).

While I was growing up, my father talked about buying a Ford Model T, just like he'd had when he was young. Finally, once all us kids were out of college, he picked up a Model A and a '31 Chevrolet. But he never forgot the Tin Lizzy. He scoffed at modern reproductions. "The pedals are wrong." "It's just got a Pinto engine in it." "Look at those modern wheels and tires."

Fig. 6-2. *Owning old airplanes is much like owning old cars. This Beech Staggerwing and Cord automobile are contemporaries.*

Finally, he bought a '23 Model T. All the running gear had been brought back to original condition, with a spiffy two-tone paint job replacing the original. "Any color you want, as long as it's black" scheme. Just what he'd wanted all those years.

As the saying goes, "Be careful what you wish for—you might get it."

In the light of 80 years of progress, the Model T is a pretty lame automobile. The brakes are mechanically operated, far less effective than the hydraulic brakes of later cars. It worked fine on dirt roads in 1935, but driving the Model T is hard work in today's stop-and-go traffic. Its suspension was designed to be robust and sturdy, not smooth. The T rode like a paint-shaker on steroids.

Starting problems had been relegated to fond memories. But now he had a *real* Model T, with fuel and spark controls, and a manual crank eager to break one more arm before being relegated to the scrap heap. It was, at least, still a Model T. Since it was so uncomfortable to drive and ride in, Dad bought a trailer and towed it to parades and car shows. On nice summer days, he cruised the neighborhood with the top down. Parts were no problem; there are a number of companies supporting the antique market. In any case, it didn't spend much time on the road. Dad had fun; he enjoyed the way it drew attention and the jaunty way it traveled down the (preferably smooth) street.

But he sold the Model T after two years and replaced it with a 1953 Ford Crown Victoria Coupe. A car with power brakes. A car that rode like a pillow in a snowfall. A car with a starter, and a heater. One he could drive to work and show off to his buddies, or take on long trips.

Antique airplane ownership really isn't much different.

Handling

Ninety-nine percent of your antiques are going to be taildraggers. And it'll be conventional gear at its worst. These planes were designed when all it took to be an aircraft designer was a piece of chalk and a hangar floor. Nowadays, a designer knows how to minimize a taildragger's quirks. Those old biplanes were designed with tail skids and grass strips in mind. Putting a wheel on the back end doesn't make them civilized.

By now, most (but not all) of the regularly-flown antiques have brakes. Often, though, they don't work very well. Too-effective brakes can contribute to noseovers, so they're sometimes deliberately kept soft. You may not have enough brake power to stop if something suddenly appears in front of you. But that's kind of moot in a lot of the old biplanes, since you won't see it anyway. As illustrated in Fig. 6-3, forward visibility is often nonexistent while on the ground. The pilot usually sits far back from the nose; on takeoff, he or she can't see the runway until the elevators become effective enough to raise the tail. Pilots of old biplanes often fly a curved final to allow them to see the runway until the last possible second.

Visibility doesn't get much better in level flight. Something blocks the view in almost every direction on a biplane. They combine the visibility drawbacks of both low- and high-wing configurations. The pilot has to keep his head bobbing to see around the obstructions. The pilot's working hard in other ways, too. Some of these

Fig. 6-3. *Periscope up! While this custom-built biplane isn't an antique, it aptly demonstrates the terrible forward visibility from the pilot's seat of many old radial-engined biplanes.*

old planes' handling characteristics are throwbacks to the bad old days. You're probably familiar with adverse yaw, where the downgoing aileron generates more drag than the upgoing one. The effect makes the plane tend to yaw in the opposite direction of the turn.

Modern airplanes compensate for it. Cessnas have Frise' type ailerons, which increase the drag caused by the downgoing aileron. Others gear the upgoing and downgoing ailerons differently to eliminate adverse yaw.

But take the famous Jenny, the Curtiss JN-4D. Designers hadn't really known about the effect when the big biplane was developed back in 1913. Pilot reports say that when the stick is moved left, adverse yaw is so bad that the *right* wing drops. Pilots must lead the turns with massive amounts of rudder.

Other flight characteristics? Well, the stall probably won't be of the gentle Piper variety. The inventor of the stall warning horn probably was still in short pants when planes were built. So, don't expect much warning. Many of these older planes will be quite happy to spin, too.

Parts and maintenance

One of the biggest impacts on the maintenance expense for antique aircraft is their out-of-production engines. Seventy years ago, the words Menasco, Ranger, OX-5, Kinner, Warner, and Hispano could be found on cowlings all across the country. Even garden-variety parts, like spark plugs have become rare. With both airframes and engines, your ability to keep these old planes running will depend upon a number of factors—*besides* money.

How rare an antique is it? Most of the spare parts for antiques are removed from junked or crashed planes. If the plane had limited production, your chance of finding a wreck for parts is greatly reduced. Like the Ford Model T, if a type is popular and many were made, parts may not be that difficult to find. The same holds true for the engines themselves; Kinner engines are popular in a number of antiques. Even if you have a rarer bird, if it mounts a Kinner you've got a better chance of keeping it running.

The question is, is it *worth* keeping it running? Nostalgia tends to hide the fact that some of these old aircraft engines weren't worth a darn when they were new. Some of those early magnetos, for instance, have extremely poor reliability. The FAA's requirement that aircraft engines have *two* ignition sources can be traced to this era.

A classic example of the questionable reliability of older engines is the Szekely three-cylinder radial, seen in Fig. 6-4. Producing 35 horsepower, it was used in a number of 30s planes, including the Curtiss Junior, the Buhl Pup, and even a model of the Piper Cub. If you see one at a Fly-In, you'll notice cables or rods circling the engine atop the cylinders. An AD note required this modification, as the Szekely had a bad habit of cracking at the cylinder bases and throwing the cylinders clear of the engine. The cable doesn't stop the cracking; it just holds the engine together so the pilot can nurse the plane down to a safe landing!

Even if your antique has a better-designed engine, you can't expect these old planes to be as reliable as their modern counterparts. Their TBOs are likely to be a quarter of that of a modern Lycoming. They'll also require more work on your part— some require greasing the rocker boxes before every flight, for instance.

Owners' groups are good sources for parts and information; they often set up an excellent network. Few require members who actually own the aircraft. Join the group before you buy; get involved with their activities. Even reticent owners tend

Fig. 6-4. *The Curtis Junior was powered by the trouble-prone Szekely radial. Note the white rods around between the cylinders; they're to keep the engine from flying apart if a cylinder breaks!*

to get more expansive when around fellow enthusiasts. Listen and learn. As Yogi Berra said, "You can hear a lot, just by listening."

Other factors

You'll spend a lot to buy your typical antique; you'll undoubtedly want to insure yourself against loss. Factor the higher premiums into your ownership equation. For that matter, you wouldn't *dare* keep an antique plane tied down outside. Or even in an open hangar, for the most part. Expect to pay more for safe storage. To make matters worse, many antiques are *big* airplanes. The ordinary General Aviation hangars available in your area may not be large enough. Moreover, moving these big airplanes around is a major undertaking, one made even worse if the hangar is a tight fit. You may have difficulty handling it on your own (Fig. 6-5). A winch may help. On the plus side, though, you'll never be lacking for willing passengers who can help you drag the plane around.

Finally, the deal kills a lot of folks: Many of these older airplanes don't have starters and must be hand-propped. It's not hard to learn, it's in fact quite easy for the aircraft using the small Continentals (Fig. 6-6), but it does tend to scare off potential buyers.

The good news

Not all the news about antiques is bad. By the mid 30s, aircraft designers were getting the hang of it. Conventional handling is more and more common to airplanes built in this period. Not all the planes are large and bulky. As mentioned earlier, the Piper Cub was in production in this era. In many cases, more modern engines have been substituted for unreliable originals. There's a Buhl Pup flying the Szekely replaced by an ordinary Continental A-65, the same engine the Champ uses.

The parts problems for antiques aren't unsolvable. Most of these planes are dead-simple—built using basic techniques and low-tech solutions. If a rare part is needed, it can usually be made. Many antiques contain a surprisingly low percentage of truly original parts, especially wood components. The use of modern epoxies and metal alloys also serves to increase the strength and reliability of these old birds.

Fig. 6-5. *Big antique biplanes are hard to handle alone. Note how the Stearman just barely fits inside an ordinary General Aviation hangar.*

Fig. 6-6. *Most antique and classic airplanes will require hand-propping. The small Continentals on planes like this Cub are relatively easy.*

And parts, especially parts for certain engines, may not be as rare as you might think. Far-sighted people started stockpiling parts years ago. Some vendors have started limited manufacture of certain critical items. Plus, several engines from the immediate prewar era had massive production runs after Pearl Harbor. Not just for airplanes, either. For instance, the 220 Continental (Fig. 6-7) radial powered some WWII tanks! True, after 60 years, these sources have started to dry up. But an overhaul of the Continental radial isn't that much different from a Bonanza engine.

While it's true that some antiques have to be babied and should only be flown occasionally, others are less delicate. A Stinson Reliant or Fairchild 24 is just as good as a Cessna 172. In most cases, you can preflight, load them up with a couple of friends and baggage, and take a long trip. *Aviation Consumer* magazine once published an ownership report on the North American T-6/SNJ (Fig. 6-8), and found that a Texan costs little more to own than the average Beech Bonanza.

Antique summary

You didn't expect it'd be easy, did you? Or even cheap? Still, there's a lot of satisfaction in owning an antique. There's camaraderie with fellow owners. There's the plain ego value of watching the parade of admirers at fly-ins.

The best suggestion, if you're interested in antique ownership, is to thoroughly research the history and availability before buying. There are a number of books detailing the design and production of certain historic aircraft. Join the clubs well before putting your money down. Talk to owners; find out about parts availability and reliability. Get known within the circles; find out who's ready to sell before

Fig. 6-7. *The Continental 220 radial engine hasn't been made in over 50 years, but thousands were built to power Sherman and Stuart tanks during World War II.*

Fig. 6-8. *Some reports indicate that planes like this T-6 Texan aren't that much more costly to own than an ordinary General Aviation retractable. However, the acquisition and insurance costs will be much higher.*

they take out advertisements. Getting known within the circle can make a big difference. Some antique owners are, well, a little possessive about their airplanes. They *care* what happens to the plane, even after they no longer own it. If you can establish yourself as a responsible person, you might get a good deal from someone looking for a good home for their pet plane. I know a case where an owner refused to sell his plane to a certain prospective buyer. The prospective buyer, it seemed, had a tape deck installed in his current airplane. The seller felt that anyone who listened to music while flying wasn't paying enough attention. He didn't want his airplane damaged by some sloppy new owner!

It might seem a strange attitude to those who have never owned an airplane. But to those who have flown or owned older airplanes, modern Cessnas and Pipers are somewhat sterile. They can't match the personality of the old-time machines. Antique aircraft, like antique cars, have a feel all of their own. If one can afford the price of admission, the cost of a flying antique is well worth it.

CASE STUDY: PARTNERS IN A BIG BIPLANE

Little did Don Connell and Stan Brown (Fig. 6-9) realize that their interest in Radio Controlled (RC) models would lead to ownership of a full-scale Boeing Stearman PT-17. They weren't just Sunday RC fliers, though. They were wrapped in the competitive world of RC Pylon Racing.

One side benefit: The races were often held on small airports. Both being licensed pilots, they got into the habit of wandering around the tiedowns and hangars after the meets were done.

One day in 1986, Stan called Don. "Wanna buy a Champ?"

$7200 later, they were airplane owners. The RC flying got put on the back burner. Connell says, "Once we were in full-scale, neither of us had time for two hobbies."

Fig. 6-9. *Stan Brown (left) and Don Connell (right).*

The Champ was nice. But the owners got the bug for a bigger airplane. "We started looking for Tiger Moths and Chipmunks," recalls Brown.

Then they flew the Champ to the annual Stearman Fly-In at Cottage Grove, Oregon. They listened to the growl of old engines all weekend. They talked to owners and discovered parts weren't much of a problems for the big Boeing biplane. Stearmans got into their blood. Brown said to Connell, "Maybe we're crazy if we *don't* buy a Stearman!"

They set themselves a $40,000 limit (this was in 1988) and grabbed a Trade-A-Plane. "We called every ad and got pictures," says Stan Brown. "I went to look at one good prospect in Texas, but the owner sold it just before I got there." They started despairing of finding a Stearman in their price range. They finally found one—in Florida, a full continent away from their homes in Seattle. They hopped an airliner to Florida and looked the plane over. It had been recently restored. Its owner had multiple aircraft and was selling Stearman N1035M just to clear some space in his hangar.

Connell and Brown bought the big biplane (Fig. 6-10) for $37,500. They then flew home via airlines and called Bill Sato, a good friend with lots of Stearman time. Bill accompanied Don to Florida and checked him out on the airplane. The two then flew the plane to Denver, where Stan met them. Sato gave Brown dual on the rest of the way home. By their triumphant arrival in Seattle, both met the insurance-company's dual requirement. They looked forward to a late-fall of open-cockpit fun.

But three weeks later, the landing gear failed. Not completely, fortunately. But 40 years of rust had critically crippled the left landing gear leg. "Any bad landing coming across the U.S. could have folded the gear," recalls Brown. Connell just shrugs. "That's what happens when you fly old airplanes."

Small consolation. They borrowed a fork lift to hoist the plane to remove the landing gear. They set it atop barrels while the gear was rebuilt. It looked like a big

Fig. 6-10. *Connell and Brown's bought their 1942 Stearman in Florida and flew it home to Washington State.*

orange Bantam brooding rusty eggs. One passerby commented: "It'll never fly, it's just an old ramp rooster."

Six weeks and $6000 later, Stearman N1035M was ready. In addition to the newly rebuilt landing gear, red, white, and blue trim graced the biplanes' nose and tail. In addition, the fuselage now sported a belligerent cartoon chicken and a brand new name: "Ramp Rooster" (Fig. 6-11).

While Connell and Brown recently sold the big orange Stearman, their experiences are typical of those who own large antique aircraft.

Ownership experience

Ramp Rooster's owners had a "handshake" partnership agreement on the airplanes, where both agreed to split the costs. Their previous joint ownership of the Champ let both men know that they could trust the other. "The AOPA partnership agreement looks good," says Connell. "I think if I went partners with anyone but Stan, I'd use it."

Once the initial problems had been taken care of, the Stearman was relatively trouble-free. "Hardly anyone wants to annual a Stearman," said Connell. "But it's getting to the point where hardly any IA wants to tackle *any* fabric airplane, including our Champ."

Once they found a willing IA, though, the process was pretty straightforward. Stearmans are dead-simple machines, with parts sized for military wear and tear.

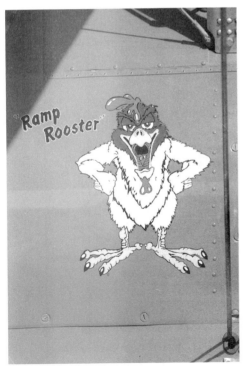

Fig. 6-11. *Colonel Sanders wouldn't have anything to do with this tough old bird. No Stearman has ever suffered in-flight structural failure.*

The 120-hours Ramp Rooster flew each year hardly showed. Connell and Brown's annuals cost $200 or so, with a ride for the IA thrown in. Note that this is much lower than even a typical 150 annual runs; an example of advantages of the informal "old boy" network that has sprung up in the antique aircraft movement.

All other maintenance was performed by the two owners, under the supervision and signoff of a local A&P. Their maintenance costs in the typical year ran about $1000. In one atypical year, Ramp Rooster's 220 Continental radial lost a rear main bearing. The partners started searching for a new engine. Luck came their way. They found a Fairchild owner who'd had a custom overhaul performed on a 220, but couldn't get an STC to install it on his airplane. The partners were able to pick up this better-than-new engine for $10,000. Add a new wooden Sensenich prop, and they were back in business.

They kept Ramp Rooster in an open hangar for most of their ownership, but eventually transitioned to a closed hangar at the same airport, paying a rent of $250 per month. Insurance (including Hull coverage) was about $1100 per year through the Experimental Aircraft Association (EAA).

For all its size and heft, the Stearman wasn't as much of a gas hog as one might assume. The Continental burned 12.5 gallons per hour—"You can set your clock by this thing," said Connell. They have an auto fuel STC. In no sense could Ramp Rooster have been considered a working machine. The majority of its 120 hours of year were local pleasure flights, with occasional trips to the annual Stearman rally in Galesburg, Illinois. Stearman N1035M made an excellent air-to-air photo platform (to which this author can attest). It was also featured skimming over the eastern Washington prairie in a video called "America by Air."

A summary of their typical costs is shown in Fig. 6-12.

Selling it

After over 14 years of ownership, Connell and Brown decided to sell most of their airplanes. Brown's mortgage business was booming, and both had developed other interests. "It got to the point where it was tough to maintain the fleet," says Stan Brown.

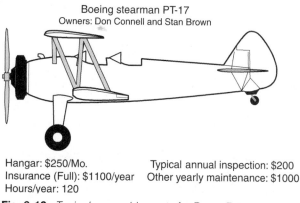

Boeing stearman PT-17
Owners: Don Connell and Stan Brown

Hangar: $250/Mo.   Typical annual inspection: $200
Insurance (Full): $1100/year   Other yearly maintenance: $1000
Hours/year: 120

Fig. 6-12. *Typical ownership costs for Ramp Rooster.*

They put the Stearman on the market—and got very few phone calls. The situation went on for about a year. Then they made one critical change to their advertisements: They listed a price. Previously, they had merely indicated "Best Offer," and received very few nibbles. But many potential buyers want to know at least the price range the sellers are looking at. They added a price to the ads, and the plane sold within a week. The buyer didn't even haggle over the price. It sold for almost double what they'd paid for it. Not a bad little investment, really. "We never did 'own' Ramp Rooster," says Connell today. "We were just the temporary curators."

Both men subsequently bought Harley-Davidson motorcycles. Connell bought a Cessna 182. Brown, an avid fisherman, started a sports fishing camp—Alaska King Salmon Adventures—and is looking into the buying his own small aerobatic biplane.

Advice

Stan Brown and Don Connell didn't fit the mold of "gentlemen" antique aircraft operators. They didn't keep Ramp Rooster in a temperature-controlled hangar on their private field. Brown's a "working stiff;" Connell is a retired engineer.

They kept Ramp Rooster flying through infusions of sweat, not bundles of cash. And through developing contacts that helped them pare the Stearman's bite on their pocketbooks.

"Some guys just come out to the airport just to fly," says Brown. "When inspections come around, they take their planes to the high-priced FBO down the street." Connell and Brown were deeply involved with the airport's social circuit. "When a guy goes out to buy an airplane," says Brown, "he becomes part of the airplane 'group'." And you can use that group to cut your cost of ownership.

For instance, their contacts found them a place that sold *retreaded* tires for their plane, at a 80 percent savings. Repairs of one minor problem were initially quoted at almost $1000, until a friend suggested an $80 alternative.

You *don't* make those contacts while renting a 172 twice a month!

"Don't wait until you're forty to start flying," suggests Connell. "Sell whatever you have to so you can buy an airplane." Brown agrees. "Find something cheap to fly, just so you get accepted by the flying crowd. The more you love airplanes, the earlier you should get into it."

CLASSICS

As we discussed earlier, Classic aircraft cover the period from the end of WWII II to 1955. An offshot is a newer category—Contemporary Classic. This category acknowledges that there were a number of aircraft in the late 50s that were still holdovers from the old days. Even the early 172s look quite different from the later models. The current cutoff is 1965.

Classics are a nice way to fly. Most of them are powered by modern aircraft engines—Continentals and Lycomings that any mechanic can maintain and find parts for. Many airframes are all metal, which precludes problems with fabric deterioration or wood rot.

Drawbacks are few, compared to antiques. As with any older piece of machinery, parts keep getting rarer. Good new parts, though, are being produced by several companies. Univair Aircraft Corporation of Aurora, Colorado, owns a number of old-aircraft type certificates. You can practically build a Stinson or Luscombe from their parts catalog.

With Wag-Aero of Lyons, Wisconsin, you can build a Piper Cub. Their "Sport Trainer" kit is a Cub replica; all of the major parts are the same as the real Piper. They also sell approved replacement parts for much of the Piper J- and PA- series.

The major drawback is that classic aircraft are more likely to be in rather sad mechanical shape. They've traditionally been the choice for those flying on a shoestring. As such, they've sat outside for 20 years and received only enough maintenance to keep them airworthy.

There's a silver lining, though. First, these are the lowest-cost planes out there. You can't really pick them up for a song anymore, but it doesn't take an entire choir. Second, the refurbishment and restoration of these older classics is becoming more and more common. A lot of mechanics and other specialists are quite familiar with the innards of these older birds. All it takes is money, of course.

But perhaps not as much as you might think. Most of the work involved in restoring these planes is "grunt" work; that almost anyone can do. Plenty of folks have restored their own aircraft. By working with a local A&P, you can perform almost all of the work involved and end up with a show winner.

Here's a good example.

CASE STUDY: FROM RAUNCHY TO RAVISHING

When you bought this book, you probably had the fond dream of finding some really cheap airplane you could buy and fly.

The thought struck Mike Furlong (Fig. 6-13) one Sunday morning in the mid 1980s while reading the classified ads in his hometown newspaper. There was a

Fig. 6-13. *Mike Furlong is a design engineer with a major aerospace company.*

Stinson 108-3 "Station Wagon" listed. Mike didn't know much about Stinsons. He hadn't thought about buying a plane at all. But the ad made him realize how tired he was of renting Cessnas. It was time to buy his own plane.

The best thing was the price: $3500!

Cheap? I bought my Cessna 150 at about the same time. I paid $6000 and thought it was a good deal. Furlong called the Stinson's owner. The plane hadn't flown for a year and a half. "Haven't been able to afford it," sighed the man. "Flew just fine, though."

Furlong drove sixty miles to the plane's base. Stinson N4095C looked a bit ratty, but then, it had been sitting without moving for almost two years. He bought it. With a little bit of work, he got the engine running. He knew better than to fly it. It didn't have a current annual, and he didn't have any taildragger time. Still, it shouldn't take much. A little clean-up, an annual, and a taildragger checkout. Nine years and $20,000 later, Mike flew the Stinson for the first time.

The resurrection of Charlie

Mike Furlong got his first inkling of trouble when he took Niner-Five Charlie to a local A&P for the annual, and the man just laughed. The fabric was shot. Mike wasn't too downhearted. He'd known the covering was marginal; he'd just hoped to get a year or two of flying out of Charlie first. He brought a trailer and started to disassemble the plane.

It took 2 hours to remove one wing's anchor bolts. They had corroded in place. Mike eventually got the plane home. The farther he looked into Charlie's innards, the worse it looked. The Stinson would require a complete rebuild. He didn't really have enough room to do the job. But his father, a long-time aviation enthusiast, became intrigued with the project. Bill Furlong built a combination shop/garage on his property, and Mike contemplated the long struggle ahead of him.

So far, he'd done just about everything wrong. He hadn't researched Stinsons. He saw the ad, liked the price, and bought the plane without an inspection. "I just got a wild hair," as he describes it. A simple prepurchase inspection would have saved him years of work and aggravation.

But from the day the resurrection of Charlie began, Mike did everything right. It started with the shop. He and his father ensured it was painted and well-lit, big enough not only for the airplane but for safe, dry storage of the removed parts. Too many people have tried to restore airplanes in single-car garages. The Furlongs went about it the right way.

Mike then bought a ton of film for his Polaroid camera. He didn't know beans about Stinsons. And once Charlie was disassembled into its smallest components, Mike knew he'd be hard-pressed to remember how to put it back together. So as he took it apart, he shot pictures. Dozens of pictures. He'd scribble on the print with a pen, highlighting the location of other components to be sure he'd know exactly how to reassemble it when the time came.

Sure, there are manuals. But they cover the big things, not how some little piece of trim might lie on the cockpit sidewall. When the time came, Mike wanted to put the plane together right. "Organization was important," he says, with considerable understatement. He didn't just throw the parts in the corner when they came off. He bought 300 zip-lock plastic bags, and stored his parts in them. All the screws for a certain panel would go into a single bag, along with a tag indicating exactly where they'd come from. Even though few would be reused, Furlong wanted to be able to tell exactly what kind of screw to buy. He filled two large cardboard boxes with plastic bags full of fasteners.

The repairs began. The rotted wood was replaced. The rusted-out steel tubing in the aft part of the fuselage was cut out and new 4130 welded in. Old parts were cleaned and reused, or new ones were found if necessary.

He eventually transferred the plane to his own house. The fuselage and tail went into the garage, but there wasn't adequate room for the wings. Only one place in the house was big enough to take the 15-foot panels—the family room. Charlie's wings stayed there for four years. "The wings kind of leaned in the corner, tucked next to the wall," says Mike.

But he couldn't maintain the pace of restoration. Things started to slow. He was getting to the point where he needed expert advice, and he couldn't really afford to hire an A&P. To top it off, he'd been having problems with his boat. He put it up for sale. A prospective buyer, an older man, came in response to the ad. He glanced into the garage.

"Stinson 108, huh?"

Turned out the visitor was a retired A&P with an IA. He offered an even swap—the boat in exchange for supervising the rest of the way on the restoration, and he'd provide the crucial signoff.

Charlie's rebirth entered the final leg. Furlong shipped the Stinson's 165-HP Franklin engine to a rebuilder specializing in these rare powerplants. He covered the aircraft with Stits fabric and painted it dark maroon with cream trim. He sent the seat covers and the interior panels to a local upholstery shop. Rather than spend time and money working on avionics, he junked the antiquated Nav-Com and installed a simple handheld radio. Mike towed the fuselage to the airport on its wheels. The wings followed on a trailer. Mike, his father, and the IA reassembled the Stinson.

Finally, the plane was ready. Furlong flew Charlie for the first time in May, 1993, with an instructor in the right seat. After 7 hours of dual, he was turned loose on his own (Fig. 6-14). Mike and Charlie flew over 80 hours in the first eight months.

Ownership experience

Stinson called the 108-3 a "Station Wagon" for good reason. It's a true four-seater, with a cavernous baggage compartment and a removable rear seat. For a total expenditure of nine years and over $24,000, Mike Furlong ended up with a practically-new Stinson 108-3 that flies faster and carries 200 pounds more than a Cessna 172.

Like most new machines, it started out with a few bugs. Oil leaks are typical in newly-restored airplanes, but that doesn't make dirty streaks on new paint any easier to take. There are always some fittings that weren't quite snugged up; in Mike's case, it was the valve covers.

In addition, Charlie's Franklin engine has the oil-filled crankshaft necessary for a constant speed propeller; plugged in this case, since Furlong installed a new fixed-pitch

Fig. 6-14. *It took more than five times the original purchase price to get "Charlie" in flying condition.*

prop. The plug leaked. Since it was right behind the prop, oil got flung everywhere. Mike eventually got the leaks stopped.

He runs 100LL fuel in his Stinson. His engine rebuilder advised against using auto fuel in Franklins, as there have been some valve problems with it. Parts are rare for the 165 horsepower engine, so Furlong intends follow the expert's instructions.

Mike's quite happy with the aircraft's performance and value, even if it did take a lot of work. "The Stinson was half the price of a used Super Cub," he explains. It gets almost the same performance as Piper's workhorse, but with a higher useful load.

Due to a long waiting list at his local airport, he originally kept the Stinson on an island airport a ferry-ride from home. After more than two years, he finally rose to the top of the list. He keeps Charlie in an open hangar that rents for $175 a month.

One financial item gave him an unusual surprise: "The insurance was lower than I expected," he says. He carries liability insurance and $20,000 of hull coverage. The first year cost $1350 a year, but he did go through some convolutions to get coverage. The first company he called required him to have 15 solo hours in the airplane before they'd cover him. The second required just 5 hours of dual and 5 solo hours. Subsequent years saw a small drop in the premium.

A pause with a partner

About two years after finishing the restoration, Mike sold a half-interest in the plane to a coworker. The author, in fact.

I had interviewed Mike for the first edition of this book and had been quite impressed with the Stinson's performance and Mike's rebuild job. I was then a member of the Fly Baby club. But when the Fly Baby was sold and the club disbanded, I was without an airplane. Mike suggested a partnership, and I took him up on it. We agreed on a value of $24,000 for the airplane, and I took out a loan on my 401K account and cut Mike a check for $12,000. We took the precaution of executing the AOPA partnership agreement, and split all costs right down the middle.

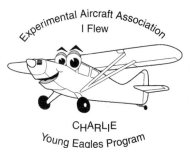

Experimental Aircraft Association
I Flew

CHARLIE
Young Eagles Program

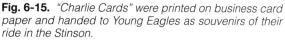

Fig. 6-15. *"Charlie Cards" were printed on business card paper and handed to Young Eagles as souvenirs of their ride in the Stinson.*

After seven years of flying a noisy, slow, open cockpit single-seater, Stinson 95C was a dream. It was about 20 mph faster, and flying in the closed cabin of the big Stinson was a lot less wearing. I took my wife flying for the first time in over ten years, and we made a couple of moderate cross-countries.

Mike and I also got deeply into the Young Eagles program. We flew dozens of kids one summer, and printed up souvenir "Charlie Cards" (Fig. 6-15) to hand out to the kids. About 18 months after our partnership began, I decided to buy another Fly Baby. Mike and I discussed how much I should pay. Mike thought I should pay less than my original $12,000, due to depreciation. But I felt I should pay more than $12,000, since the classic Stinson's value had probably risen a bit. We compromised on the original $12,000, and Mike was back to single ownership.

Maintenance lessons

Mike's intention from the start was to participate in all the annual inspections—after all, who knew the airplane better? His annual inspection costs initially ran between $500 and $1,000. But about three years after we'd dissolved the partnership, he became involved in another major project: He and his wife decided to remodel their home.

Like the Stinson, the little project turned into a great big one. It became so involved that Mike quit flying for years. He drained Charlie's oil and replaced it with heavy storage oil. It has a strong clinging effect; it tends to coat the engine parts and delay corrosion. He ran the engine every six months, but otherwise, Charlie sat idle in the hangar.

After five years, Mike felt Charlie had been inactive long enough. He was ready to not only take it out of storage, but to add some modifications—new gyros, refurbish the propeller, add a spin-on remote oil filter to the engine, replace a sticky fuel valve, replace the transponder, install a panel-mounted comm radio, and so on. While the house project was nearing its end, he was still heavily involved in it. So instead of taking on another Charlie rebuild, he turned the plane over to the local FBO.

It did produce a few problems. The main one is that the FBO's mechanics, while competent, had next to no experience with classics or antiques. They were more used to routine maintenance of contemporary Cessnas and Pipers—the Stinson company had put their planes together quite differently! It took a long time and quite a bit of money before Charlie was operational again. The total bill exceeded $12,000.

Advice

Mike had known Charlie had problems when he bought the Stinson. He expected some amount of work would be necessary. After all, the plane had been flying just 18 months earlier. He did end up with a gorgeous Stinson. But if he had known the plane's true condition, he would have passed it by.

He'd been a little too enthusiastic at the prospect of owning his own plane. "Sleep on it," is Mike Furlong's simple advice to the prospective aircraft buyer. "Get a pre-purchase inspection."

Figure 6-16 summarizes Mike's cost of ownership.

Stinson 108-3 Station Wagon
Owner: Mike Furlong

Hangar: $175/Mo.   Typical annual inspection: $1000
Insurance (Full): $1350/year   Other yearly maintenance: $500
Hours/year: 80[*]

[*]Aircraft recently emerged from 5-year inactive period

Fig. 6-16. *A summary of Furlong's owerhship expenses.*

7

Special Wings, Part 2: Homebuilts

When the General Aviation industry went into a slump in the 1980s and 1990s, one type of airplane took off in leaps and bounds: Amateur-Built aircraft—"Homebuilts" to their friends. Since the early 1980s, interest in homebuilt aircraft has grown from a flickering campfire to a roaring inferno; from low-and-slow fun airplanes to 300 mph fiberglass speedsters (Fig. 7-1). The Experimental Aircraft Association's (EAA) annual fly-in convention at Oshkosh is the world's largest aviation show.

Every year, new homebuilt designs vie for the attention of a rising number of people intent on building their own plane. A number of companies have sold over a thousand kits, each. And that's kits, costing tens of thousands of dollars each, not just plans. One company has over 4000 of its plans and kit-built designs flying! But there's more to a homebuilt than signing a check and spending a couple of weekends in the garage.

HOMEBUILT BASICS

To begin, let's establish what, exactly, a homebuilt is. Back when you were taking flying lessons, your instructor showed you a little plastic pouch in the back of the airplane. Inside were a couple of slips of paper. "You gotta have these on board to be legal," your instructor said. "Look, here's the registration, radio license, and the airworthiness certificate." The instructor then stuffed the forms back in the packet. You probably never looked at them again. Two of them cover mere bureaucratic formalities. The Registration verifies that the FAA knows about your particular airplane, and the station license authorizes you to use the Com radios.

But the Airworthiness Certificate—now there's a story.

The factory that built your trainer spent thousands of man-hours to be able to put that little scrap of paper in your airplane. To earn certification in the Normal,

Fig. 7-1. *In 25 years, homebuilts went from wood-and-fabric sport machines to fiberglass speedsters.*

Utility, or Aerobatic categories, the manufacturer proved the plane could withstand certain G-forces, and possessed certain safety and warning equipment. Its flight characteristics had to meet stringent government requirements. Its engine came from the narrow ranks of FAA-approved powerplants. The government sent its own test pilots to verify the findings of the company.

The reward came when the FAA awarded the design a Type Certificate (TC). The Airworthiness Certificate crammed into that flimsy holder guarantees that this one particular plane met all the requirements of the TC.

The next time you're around that little plastic packet, take out the certificate. Smooth the wrinkles; unfold the dog-eared corners. Many of our fellow pilots and dozens of my fellow engineers worked hard to put that tiny piece of paper in your plane.

The experimental category

All that effort is expensive. It can only economically be put into production aircraft, where the cost can be amortized over hundreds of planes. But the FAA recognizes that certain limited-purpose airplanes don't really require such extensive analysis and testing. For these aircraft, the FAA provides the "Experimental" category. Aircraft with Experimental certificates don't have to prove anything before they're allowed to fly. The FAA only demands that precautions be taken to protect the lives and property of those not operating the aircraft.

The Experimental category includes a number of subcategories, including "racing," "research and development," and "amateur-built." Of these, amateur-built has the most relaxed rules. To qualify for the amateur-built category, at least 51 percent of the labor, required to build the plane, must be done by a nonprofessional builder "for educational or recreational purposes."

Homebuilt advantages

Aircraft design, production, or homebuilt, is a study in compromise. The designer merges the performance objectives, FAA rules, customer preference, and marketing inputs to generate an aircraft design. He or she cannot hope to please everybody; rather, the designer's goal is an aircraft that most can accept with minimal grumbling.

But experimentals aren't bound by the FAA's conservative design requirements. The design can be optimized toward any specific goal the designer wants. With the ability to optimize the aircraft toward a particular goal, the homebuilt-aircraft designer can produce aircraft, which outperform production planes, and seriously attract attention at the same time (Fig. 7-2).

Speed is a typical homebuilt design goal. Take the Lycoming O-235 engine mounted in the Cessna 152. Several homebuilt designs use this engine; one cruises 100 mph faster than the 152!

There's no magic involved. That particular homebuilt has over a quarter less wing area. It has retractable gear. The occupants sit semireclined; the aircraft's drag is but a fraction of the Cessna trainer's. Homebuilts cost less to operate, as well. If you owned a 152, you'd be spending money every year for a mechanic to perform the annual inspection. And if something broke, you'd have to replace it with parts approved for aircraft use.

Not so with a homebuilt. If you built it, the FAA will give you a Repairman Certificate. This authorizes you to perform all the maintenance and inspections on your bird. You can replace a broken component with anything you feel is safe. You can even use a converted car engine and avoid high overhaul costs. Acquisition costs are cheaper than that of a new production bird. A Lancair IV might cost $200,000 by the time it's

Fig. 7-2. *Homebuilts like this Velocity get better performance on the same engines as production aircraft—and certainly arrest the attention. (Courtesy Velocity Aircraft.)*

finished and fully equipped, but its performance exceeds that of a $600,000 Bonanza. Plus, you could buy the components of that kit gradually over time, stretching out the cost over a number of years. This has the effect of a no-interest loan—sort of a lay-away/fly-away plan.

Finally, production aircraft are, well, rather boring. They're the equivalent of four-door Chevy sedans. A Cessna 172 or Piper Warrior might be good transportation, but one can't get too excited about them. Neither does anyone else. Ever heard the term "JAFCO" around the airport? It stands for "Just Another Friendly Cessna One-Seventy-Two," or words to that effect.

Homebuilts are the sports cars of the air. Their controls are light and they're fun to throw around the sky. There's a lot of satisfaction involved in building your own plane, and you'll be sure to draw a lot of attention.

Homebuilt disadvantages

When compared to new factory aircraft, the cost of building a homebuilt looks like a real good deal. However, the homebuilt's biggest competitors are used certified aircraft, and the homebuilt often comes out second-best.

You can build an IFR Glasair II or Lancair Legacy composite speedsters for $60,000-$80,000, and cruise about 200 mph. For the same money, you could buy an early 70s Mooney and carry two more people, only a bit slower. And be flying tomorrow, not five years from now.

It's the same for aircraft on the other end of the speed scale. The $30,000 or so you'll spend to build a Skystar Kitfox could pick up a nice Champ. And make no mistake about it: It doesn't matter if the plane comes as a "kit," building an airplane is a lot of work. Don't be misled by the construction estimates from the kitplane companies. In the real-world, builders often double or triple the time the kit manufacturer's advertised construction time.

You'll also need somewhere to build the plane. Homebuilts have been constructed in apartments and single-car garages (Fig. 7-3), but it's certainly not the optimal course. There'll be tools to buy, too. There's another worry: Once your plane is registered, you are the legal manufacturer. If you sell the plane, and the buyer (or a subsequent buyer) crashes it, you might be sued.

Finally, homebuilts have come under a bit of controversy regarding their handling characteristics. Remember, aircraft design is the art of compromise. Designers of production aircraft usually sacrifice performance in the interest of meeting the FAA's handling and stability requirements. There is no free ride in aircraft design. If you push one side of the envelope out, another side squeezes in.

Let's look at an example in the production-plane world. Take your humble O-235-powered Cessna 152. The same engine was installed in an earlier aircraft, the American Yankee. The Yankee cruised 10 mph faster and had crisp, sporty handling. It also has one of the highest accident rates of the General Aviation fleet. It's fatality rate is *three times* that of the Cessna 150. The Yankee has been particularly faulted for its poor stall characteristics.

So what happens when you push the speed further? The best illustration comes from Australia, where homebuilt designs must be flight tested and approved.

Fig. 7-3. *Finding a place to build an airplane is the first hurdle. This Osprey amphibian is being built in a single-car garage. Note the jackstand in the doorway to allow part of the door jamb to be removed. This is the only way the plane can come into and out of the garage.*

One of the United States' most popular homebuilts flunked. And we aren't talking about some funny business in a seldom-seen corner of the flight envelope. When an experienced Australian test pilot stalled the aircraft, it flicked inverted. A thousand feet were lost in the recovery, with a professional test pilot at the controls. Several other popular U.S. designs have just barely passed muster down under.

However, don't be frightened away. What it means is that homebuilts may fly differently than the production planes you're used to. They'll be a little more challenging to fly precisely, but most of them will be safe. Take that American Yankee, for instance. A deathtrap? Not if you take a few hours of lessons from an instructor well-versed in the Yankee's idiosyncrasies. I did, and found the Yankee a joy to fling around the sky. For fun flying, I'd pick the American (or one of its Grumman descendants) over a 152.

Interestingly enough, Yankee was originally designed as a homebuilt!

Should you build?

Should you build a homebuilt?

Homebuilt aircraft accident rates are about 25 percent higher than production aircraft, but the accident causes tend toward builder and maintainer error, not faulty design. Structural failure is almost invariably caused by deviations from the plans or exceeding the aircraft's flight envelope. But it's obviously important to build it right.

Fig. 7-4. *Buying an airplane takes just a day—building one takes years. Are you sure you want to wait?*

As far as stability and handling, research the aircraft before you pick a kit; get a good checkout before your first flight. Homebuilts aren't difficult, only different. But it still takes a lot of effort to get to that first takeoff.

The question is, do you want to *build* a homebuilt? Not just own one. Not just fly one. But do you actually want to go through the years of effort (Fig. 7-4) required to construct your own airplane? Because if you're looking at the construction process merely as an intermediate step to flight, you'll probably fail.

If you want to fly a plane, *buy* a plane. If you want to *build* a plane, then build one. It's as simple as that. If you really like the designs but decide that you really don't want to build one, take heart. Later in this chapter, we'll discuss buying used and partially-completed homebuilts.

HOMEBUILT CONSTRUCTION

Can *you* build your own airplane?

Probably. In the early days, homebuilders mostly came from the ranks of those with considerable shop skills—folks who were willing to spend thousands of hours building a simple single-seat sportplane. Today's kitplane era changed all that. No longer is welding considered a vital homebuilding skill; nearly all kits come with prewelded parts. No longer do builders have to laboriously assemble wooden ribs or shape metal ribs with a bar of solder; most kits supply ready-to-install ribs.

Of course, that doesn't make building easy. There's still a lot of work to do and new skills to learn. In this section, we'll discuss some of the aspects of construction.

Type of construction

One factor to include when considering homebuilts, is selecting which type of construction you'd like to work with. As Fig. 7-5 illustrates, there are four basic types:

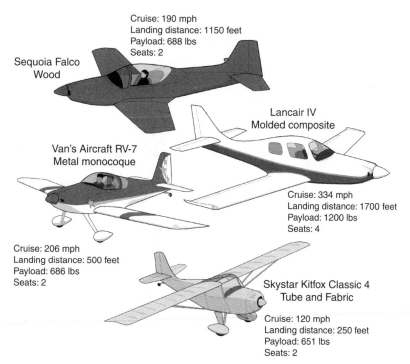

Cruise: 190 mph
Landing distance: 1150 feet
Payload: 688 lbs
Seats: 2

Sequoia Falco
Wood

Lancair IV
Molded composite

Van's Aircraft RV-7
Metal monocoque

Cruise: 334 mph
Landing distance: 1700 feet
Payload: 1200 lbs
Seats: 4

Cruise: 206 mph
Landing distance: 500 feet
Payload: 686 lbs
Seats: 2

Skystar Kitfox Classic 4
Tube and Fabric

Cruise: 120 mph
Landing distance: 250 feet
Payload: 651 lbs
Seats: 2

Fig. 7-5. *Construction methods used in homebuilt aircraft.*

1. Composite
2. Metal monocoque
3. Tube and fabric
4. Wood

Composite construction works by combining two (or more) materials whose advantages complement each other and whose deficiencies cancel out. In the simplest form, strong-but-flexible fiberglass cloth is soaked in stiff-but-brittle epoxy (itself a mixture of resin and hardener) to form strong-and-stiff components like cowlings and wheel pants.

The two types of aircraft composite construction are molded and moldless. The first uses a mold to define the shape of the structure. It's most suited to a production line. The kit purchaser bonds the parts together just like a plastic model. With moldless composite construction, on the other hand, the builder shapes the design from Styrofoam and fiberglasses it. It's more time-consuming, but cheaper.

Metal monocoque construction is used by aircraft manufacturers from Piper to Boeing. Take a thin sheet of flat metal—it's pretty flimsy. But roll the metal into a wide tube and it becomes stiff. Attach some bulkheads inside to prevent the tube from collapsing, and you've got a light, strong structure ideal for aircraft (Fig. 7-6).

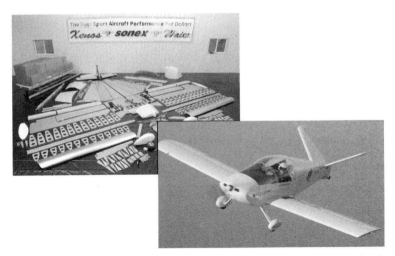

Fig. 7-6. *Much of the aluminum used in the strong but light Sonex series is around twenty-five thousandths of an inch thick. Yet it withstands over six Gees. (Courtesy of Sonex Ltd.)*

Tube-and-fabric construction is another of the "traditional" ways to make light aircraft. The main structural shape of the fuselage is defined by a metal truss. The wing can be built using several methods. The spars, for example, can be solid wood, built-up wooden boxes, or metal extrusion. The ribs can be stamped metal, cut plywood, built-up shapes, or even foam. Fabric is applied to the structure (hence the term "ragwings"), then sealed with dope to produce an enclosed, streamlined shape. Commercial tube-and-fabric lightplanes include classic Pipers and the Bellanca Citabria.

Wooden aircraft are built just like balsa models. The fuselage structure consists of longerons and bulkheads glued into the proper shape. Like tube-and-fabric aircraft, the wings can be built in several ways and are either sheathed in plywood or covered with fabric.

The preceding is only a general guide. A number of airplanes use a combination of methods. Generally, though, designers keep to one mode of construction throughout. Not only does it make the parts list simpler, but it reduces the construction time, since the builder won't have to learn two unrelated skills.

Types of kits

The cheapest way to build an airplane is to buy a set of plans and build it entirely from scratch. That route isn't all that common anymore. The kit homebuilt, or *kitplane*, has taken over much of the homebuilt arena.

There are some good reasons. First, while a kitplane will invariably cost more to build, the basic price is at least known in advance. There's a little less uncertainty. Second, kit companies usually have good builder support. Large manufacturers provide a help-line or special email support for builders. Third, a successful kit builds further interest. If a prospective builder is wondering if an ordinary person can build a plane, a line of VanGrunsven RV-7s and RV-8s at a local fly-in is a pretty powerful argument.

All kits are not created equal, though. The amount you pay depends upon how much work is done for you. The materials kit contains the raw materials necessary for a plans-built aircraft. No work is performed by the kit supplier; in fact, the supplier often has no connection with the aircraft designer. Basically, it's a painless way of obtaining the necessary hardware. The kit provides aluminum sheets, plywood, metal extrusions, and other materials which require considerable work to change to aircraft parts (Fig. 7-7).

The next step upward are those designs, which can be built from plans, but offer subkits to hasten construction. This is the most flexible method—the builder can buy whatever subkits his or her budget allows, and build the rest of the airplane from scratch (Fig. 7-8). Or a builder unsure of his skills can buy subkits for critical or complex components. Of course, when the entire aircraft can be ordered as a single item, it becomes the last category; a complete kit (Fig. 7-9). Few are truly complete; even the best don't include paint or batteries. The ultimate complete kit is the "Quick Build" kit, where the kit is delivered with significant portions of the basic structure completed.

Construction cost

The materials kit, as mentioned earlier, is the cheapest because nothing is done for you. The most expensive kits are usually those that are for molded composite aircraft. It isn't because the fiberglass and epoxy are that costly; rather, the buyer pays for the factory labor necessary to make the parts. However, the cost of the kit isn't the only factor involved. No kits include everything required to build the plane. There's some additional hardware to buy; from penny-ante stuff to big-ticket items.

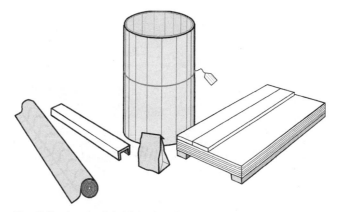

Fig. 7-7. *A materials kit is the cheapest way to get into home-building. But it leaves you the most work!*

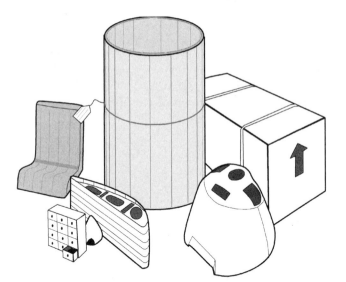

Fig. 7-8. *Building aircraft under the subkit method gives the builder the most flexibility. He or she can save money by doing most of the work, or buy prefabricated components.*

The biggest of the big-ticket items, of course, is the engine. Currently, only designs using low-cost two-stroke engines include the powerplant. For planes using conventional aircraft engines, the additional cost is frightening. Engine prices can range from a few hundred dollars for a used two-stroke to well over $60,000 for the larger engines. The kit for the Van's Aircraft RV-7 sells for about $18,000, and includes "just about everything needed except the engine, prop, instruments, and tires."

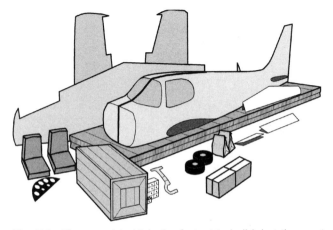

Fig. 7-9. *The complete kit is the fastest to build, but the most expensive.*

A new Lycoming O-320 engine will cost about $25,000.

The rest of the additional items aren't quite as bad. Propeller? Two hundred and fifty dollars for a Rotax prop; thousands for a constant-speed model for a Continental or Lycoming. Instruments? At least $800 for basic VFR, if you're buying new. Avionics? No kit includes radios or a transponder. Simple VFR birds can get by with just a handheld transceiver, but these sell for $250 and up. By the time you add GPS and a transponder, avionics cost is well past $3000.

That's not all, either. How well equipped is your workshop (Fig. 7-10)? Whatever tools you own, they probably aren't enough. Drill presses, band saws, table saws, grinders, it all adds up. Some kits claim that they "Can be built with simple hand tools," but you'd be better-off getting a drill press and an aluminum-cutting band saw, at least.

Construction time

All kit companies include a typical construction time with their advertising. Most builders laugh at these estimates.

Generally speaking, you should double (or triple) the manufacturer's prediction. If you want to spend the least time building, pick less-complex aircraft. The RV-7, for instance, has a simple, rugged, fixed gear. Most models of the Lancair have a mechanically-operated retractable tricycle system. The time a Lancair builder gains through composite construction might well be lost when it comes to gear installation.

Of course, the kitplane manufacturer can reduce the effect of additional complexity by including extensive prefabricated components. If an assembled gear

Fig. 7-10. *An additional cost when building your own plane is the outlay for the tools.*

linkage can be supplied, the time impact of retractable gear is reduced. However, there is a constant factor in all kits, one that sets the rock-minimum time the kit might take to build. Whether, you're building a Lancair or a Kitfox, an RV or a RANS S-7, you'll spend most of your time working on the internal systems. All of them have controls systems to install and trouble-shoot; all have wheels and brakes to install; all must be painted. It's fun to build primary structure; to make something that *looks* like an airplane.

But most of the time, you're standing on your head in the cockpit or crouching in front of the firewall, working on some subsystem or another. Figure 7-11 shows some of the systems, which must be built and tested on most homebuilts.

Construction summary

Most kits proudly state, "Can be built with ordinary tools by someone with average skills."

What's an ordinary tool? What's an average skill level?

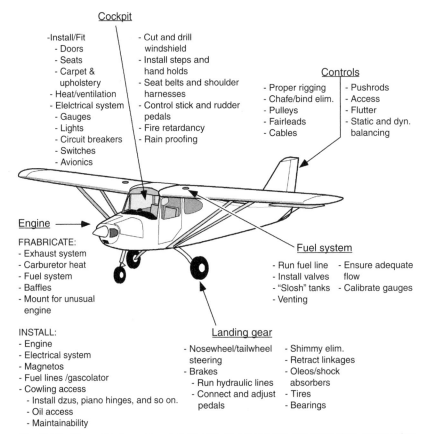

Fig. 7-11. *No matter how complete the kit, all will require extensive time-consuming subsystems work.*

The definition changes between kit manufacturers. You can be sure by the time you're done, you'll have much higher skills, and a much better-equipped shop. It all comes at a price, of course. Both financially and personally.

It's hard to answer all your questions in these short pages. I've written another book, *Kit Airplane Construction*, for those contemplating building a kit aircraft. The book covers the selection process, as well as basic workmanship techniques for all four types of construction.

BUYING USED HOMEBUILTS

If you can't face the construction process, but the appearance or performance of a particular homebuilt is still attractive, consider buying a used homebuilt. Prices can be quite reasonable. Some can be bought for little more than the total cost of construction.

There are a number of reasons. First, some folks like the building process more than flying. Some have built five or more aircraft, and quite often are craftsmen with excellent workmanship. Second, the occasional builder doesn't do his homework before starting construction, and the resultant airplane doesn't meet his needs. Finally, a builder could own his creation long enough to get tired of it and wish to move on to something else (Fig. 7-12).

Aviation writer Budd Davisson has a saying: "You're better off buying a used snake than a used homebuilt. . . ." Obviously, significant problems could exist with the aircraft. The structure is already closed up, making detailed inspection difficult. It's quite possible for number of hours to be flown without a hidden flaw making itself known. In one horrible example, a homebuilt flew 14 hours without wing bolts. The mistake was discovered when the wing separated in flight.

Inspections

A prepurchase inspection is even more vital for a used homebuilt than a used production aircraft. Find someone with experience on the type of aircraft, preferably

Fig. 7-12. *After flying for a while after completion, some builders get bored with basic airplanes like this Pietenpol, and put them up for sale.*

someone who has built the same model. They should pass judgment on the workmanship and adherence to the construction plans. While some deviations are minor, changes to the control system, rigging, or basic structure should set the alarm bells ringing.

An inspection by an A&P would also be a good idea, as they're experienced in rapid assessment of airframe and engine condition. However, for Rotax-powered aircraft like the Avid Flyer, check at the local ultralight center for someone who knows 2-stroke engines. The aircraft should be test-flown by yourself and the experienced builder you brought. You can judge if the plane is right for you; the experienced builder will determine if the plane flies like it should. Examine the logbooks. An aircraft that is flown regularly probably flies well (Fig. 7-13); a plane that sits a lot, might have problems.

Single-seat airplanes present a problem. Will the owner let you test fly it? There are a number of cases of prospective buyers crashing single-seaters. If your experienced builder-friend arrives in his own version of the same model aircraft, the owner shouldn't have too many worries. Or, if the aircraft is otherwise acceptable, you may be able to buy the aircraft contingent on an acceptable test flight.

Maintenance

Maintenance is another issue. You cannot receive a repairman certificate; you didn't build the aircraft. The original builder can perform the maintenance, if he still retains his certificate and is willing. Neither is guaranteed. If buying the plane from the original builder, a fresh annual inspection should be part of the deal. Have an outside mechanic inspect the aircraft as well, but licensed mechanics charge more for annuals than prepurchase inspections.

Don't fault the builder if he or she is unwilling to continue maintaining the aircraft after you buy it. The builder probably isn't a professional mechanic; the legal

Fig. 7-13. *If the owner flies the airplane a lot, it probably doesn't have any handling or safety problems.*

and other repulsing pressures are strong. If you buy from a friend, though, they may be willing to continue. Any A&P mechanic is allowed to maintain and inspect homebuilt aircraft. One difference between homebuilts and production aircraft is that the A&P doesn't have to be an IA to perform an annual on a homebuilt. That'll save a bit. But if the plane has a Rotax or converted auto engine (Fig. 7-14), it might be hard to find an A&P willing to be legally responsible for its maintenance and continued operation. Check around before you buy the aircraft.

One interesting point is a gentleman's agreement between the FAA and the EAA. The owner of a homebuilt aircraft is allowed to maintain his airplane, as long as an A&P signs off the work within the next 12 months. If you can find a mechanic willing to work on this basis, you've got the best of both worlds—authorizations equivalent to a repairman certificate, with an experienced eye to keep the aircraft safe.

The final analysis

As you're discussing the deal with the seller, one of the primary things you should determine is, *why* the aircraft is for sale. Does the plane just have a few hours on it? That should trigger a few red lights. Maybe its handling is so weird that the owner is scared of it. Maybe he or she discovered some major mistakes made during construction.

Examine the aircraft logs. Has the plane flown regularly? If the plane were 10 years old, I'd be a lot happier to see a thousand hours in its logbook rather than one hundred. Good-handling; reliable aircraft are flown. A plane airborne for only a few hours per year might have problems.

Probably your safest route is to know the seller in advance. I've seen a number of homebuilts come up for sale within my local EAA chapters. With my knowledge

Fig. 7-14. *If you're going to buy a homebuilt with a converted auto engine, make sure you're prepared to keep a unique power system operational.*

of the builders, I would have had no hesitation about buying most of the aircraft. With the boom in kitplanes, we are going to see more and more homebuilts available on the used market. They're a definite option for those looking for something more than a typical factory plane, but without the patience or inclination to build their own.

BUYING PARTIALLY COMPLETED HOMEBUILTS

There's a thriving market in flyable homebuilts. Uncompleted kits are another kettle of worms. Buying and selling unfinished homebuilts is nothing new. It used to be said that 30 percent of homebuilt aircraft projects were eventually flown—10 percent by the first builder, 10 percent by the second owner, and 10 percent by the third or subsequent owners. One Fly Baby ran through six owners until it was completed, 20 years after construction started.

A flying homebuilt has passed its most critical test, and one can use its flying characteristics to judge how well it's built. An uncompleted project? Who knows what might be wrong with it? The sixth owner of that uncompleted Fly Baby found a critical problem with the wing bracing system. The *sixth* owner. Would the previous owner have detected the problem before it was too late? Still, there are some advantages to buying partially-completed planes. The primary one is the work you won't have to do. Wouldn't you like to pick up a completed empennage for the cost of the tail kit alone?

In addition, buying an uncompleted plans-type homebuilt can be the equivalent of buying a kit. Like to build a Volksplane (Fig. 7-15)? You might be able to pick up most of an airframe for less than a thousand dollars, and get it flying for a couple of hundred hours and just a little more money.

Fig. 7-15. *Not all builders finish the airplanes they start. If you like Volksplanes, buying a partially-completed example can give you a head start.*

Buying a partially completed homebuilt is a gamble. It might hide a deadly hidden flaw, or it might be the best aircraft deal since the $650 Jenny. The problem, of course, is telling the difference. How can you make sure you're not getting taken?

Do your research

To begin with, know what you're buying. Check the specifications. Read pilot reports. If the aircraft type never became popular, there's often a good reason—usually rooted in flying qualities or being difficult to build. Avoid partially-completed custom designs; stick with those which have been commercially available.

If the advertised project is a kit, knowing the exact model is vital. Homebuilts have drastically improved in the last 10 years; an older kit's performance and reliability are often reduced compared to the latest model. Once you understand the aircraft, find someone who's built one. Ask about common problems. What mistakes are typical? Which are the critical operations? Ask them if they know the builder with the project for sale; get their impression of the owner's workmanship. Of course, most folks will be reluctant to slur an acquaintance's ability. Rave reviews are a good sign; a shrug and a "pretty good" can mean anything.

The examination

Thus armed, it's time to examine the merchandise. Don't go alone. Four eyes are better than two, and if the second set belongs to a builder experienced with the design, so much the better.

Scope out the builder's shop—how much equipment was he working with? Did he have an adequate air compressor to rivet with, or some cobbled-up system using a hobby compressor and a small air tank? A sturdy workbench or a cheap folding table? How did he store the kit materials—did he lean plywood against the walls, or build a flat-storage rack? Just a few discount-store hand tools or a complete machinist's set? This isn't to say that one can't build an acceptable aircraft with minimal tools and facilities. But you're looking for clues toward poor workmanship. Few homebuilders rush out and completely equip their shop from the outset. Someone with a well-equipped shop probably owned the tools prior to beginning the aircraft; their workmanship would probably be good from the start. But if the owner has a less-than-optimal compressor setup, his rivets may have suffered. A flimsy worktable can result in warped components. Poor storage of plywood, steel, or sheet aluminum can ruin good material.

Speaking of the raw materials, check them over for damage and decay. For wood, look for the discoloration of rot and the flaking, and checking due to drying-out. The end grain of wood stock should have a coating of paint or varnish to retard the drying process. Aluminum should be free of corrosion. Look for roughness and dry powdery areas. Sheets often get lightly scratched in normal shop handling, but there is a limit. Deep scratches concentrate stresses and can cause premature failure. Slight bends and wrinkles may not affect the strength, but make it awkward to work with.

Any untreated steel is likely to have a patina of rust. Conscientious builders apply a coat of paint as soon as possible. Unlike corrosion on aluminum, rust can be safely removed from steel, as long as it isn't deep. If the builder has neglected to prime a steel-tube fuselage, though, de-rusting it will take considerable time.

You won't have these worries with composite materials. However, epoxies usually have a maximum shelf-life. Make sure it hasn't been exceeded. Some epoxies are shipped in slightly permeable plastic containers. These reduce the shelf life, as well as stink up the shop. It's a positive sign if the builder had transferred the resin to metal or glass containers. Finally, make sure any fiberglass cloth is dry, clean, and still on the roll.

Checking the aircraft

Of course, your primary item of interest is the aircraft itself. Check its straightness—crouch down and squint along edges. Run your hand along surfaces and check for smoothness. Have the builder remove access panels and other items that hinder a full inspection. You'll soon understand one of the biggest frustrations in buying a partially-completed project: The more finished, the harder to inspect. A closed-up metal or composite wing may cover a number of sins, yet the buyer will want a higher price due to the level of completion.

Specific areas of interest vary with the type of construction. For wood, pay careful attention to the glue joints. Find out what type of glue the builder used. The poor filling qualities of traditional glues require tight joints for strength. Modern epoxies such as T-88 do a good job of filling loose joints, but tight ones are still a sign of good workmanship. One critical item to check is the amount of glue in the joints. They should all show signs of glue oozing out when the pieces were clamped. The excess is normally wiped away, but a small fillet should be visible all the way around.

Similarly, composite layups should also be checked for starvation. Completed parts should have a rich amber color; pale patches indicate voids and delamination. An excess of epoxy isn't dangerous, but it's heavy and a sign of sloppy workmanship. The cloth weave should run straight or curve evenly around bends. All parts should be dry and clean of grease and dirt.

Partially-completed aluminum aircraft requires careful examination of the riveting. Examine the skin around the manufactured heads—look for distortion, dents, and other signs of improper procedure. Check both heads for distortion or cracking. Gauge a few shop heads, make sure they're at least half the rivet diameter high, and 1.5 the diameter wide. Mind the edge margin—the rivets should be spaced at least three times their diameter apart, and at least twice the diameter from any edge. Check the straightness of the rivet lines. A little waviness is normal (and for some of us, unavoidable), but you can get a better feel for the builder's level of workmanship. Look for scratches and dents.

Welded structures can be tricky. Factory-welded areas are generally all right, although I've seen a few bad ones. Builder welds are another matter. While it's easy enough to check the surface, it's quite possible that only the surface material had been melted, and that it has not combined with the underlying metal. "Penetration" of the weld is important, but difficult to prove without actually cutting through the joint. Hopefully, the builder made periodic test welds to saw apart and verify penetration.

Other goodies

In addition to the aircraft itself, don't forget to examine any other equipment included in the sale. Checking an engine can be difficult, but at least make sure it includes the required accessories, such as magnetos, alternator, starter, and carburetor (Fig. 7-16). Additional firewall-forward goodies sweeten the deal, especially an engine mount.

If the engine's used, scan through the logs. The engine should have been "pickled" for storage. Special oil coats the crankcase and sump, desiccator plugs are installed in the cylinders, exhaust, and breather, and the engine should be wrapped and stored in a dry place. Otherwise, it could be just a 250-pound block of rust.

Check the plans, make sure they're complete. Glance over the inventory of kit components. If the kit originally included parts such as wheels, tires, and instruments, find out if the owner has included them in the sale. Ask to see the owner's builder's log—the more detail, the better. For example, it should list exactly what brand of primer was used. That'll help you avoid compatibility problems when it comes time to paint. If he's got a collection of newsletters or similar publications, ensure they're included in the deal. Ask the owner about construction variations. Most builders make small modifications; few of these affect complexity or airworthiness. But be leery of major mods, like conversion to a nonstandard auto engine or a scratch-built retractable gear. There's nothing wrong with people making changes like

Fig. 7-16. *If you buy a partially-completed homebuilt that comes with an engine, make sure the powerplant includes all components necessary for operation.*

this; however, you don't want to be stuck with someone else's poorly-thought-out modification.

The decision

Finally comes decision time. Has the builder done a safe job, so far? Homebuilts can be remarkably tolerant of bad workmanship. As long as the structure is sound, the cosmetics don't affect the airworthiness. But it's your peace of mind as well as your life at stake. It's no fun flying a plane you have to worry about all the time. And everyone's going to assume those bad cosmetics are your fault.

If you decide the kit is acceptable, it's time to haggle over price. Unfortunately, there are no guidelines; no "blue book" for unfinished kits. Their value is determined by too many factors, from the workmanship to completeness to the kit's popularity.

Don't be thrown off by the "current" kit price. The original Glasair kit sold for half today's price, but had poorer flight characteristics and was markedly harder to build. So, the starting point should be around the original cost, not what a brand-new kit might run.

How much to pay for the work already completed is going to be the thorniest part. If you have to redo shoddy work, the project is of less value, though it might not be politic to tell that to the owner. It boils down to how much the work is worth to you. Treat the seller's description of "X% completed" with a ton of salt.

For a starting point, take the original kit price (or cost of the materials), then add deltas up or down depending upon workmanship and additional equipment. The actual market value is probably between 80 and 120 percent of this amount.

What to offer? Depends on the situation. Some sellers are just tired of the bother and are willing to take a much lower price. Others are in no hurry and are willing to wait for what they feel is a fair price. I know someone who sold a project, then bought it back later for half what he originally sold it for. Each case is different. It never hurts to make a low offer and see how owner reacts. But unless you pay peanuts, it's probably less of a deal than you think. Talking to local buyers of partially-completed homebuilts reveals one common theme: They had to do more work than they thought would be necessary. As everyone will tell you, the last 10 percent of the aircraft takes 90 percent of the time.

Finally, remember that you must perform at least 51 percent of the work on the aircraft to receive a repairman certificate. A partially completed project may not allow you to qualify to do your own maintenance. Get a ruling from an FAA inspector prior to buying an unfinished project.

CASE STUDY: A BRIGHT GREEN GEM

Rich Shankland didn't start out planning to build his own airplane. The Seattle firefighter's ownership of a classic Ercoupe in the late 1970s ended when he was stranded away from home by a hole in the piston. Normally a sour note—except he sold the plane, as it sat, for more than he paid for it! He didn't replace the Ercoupe with another plane, and eventually retired to a home on the West side of Washington's Puget Sound.

Then he received a curious inheritance. His daughter's father-in-law passed away, and in his effects were the plans and some raw material for a two-seat home-built. The man had bought the plans 25 years earlier, but after an abortive start, he never got back to the project. As no one else in the family was interested, Rich became the inheritor of the plans and parts.

The plans were for the Piel Emeraude. In the late 1950s, Frenchman Claude Piel had designed a pretty little two-seat homebuilt he called the Emerald—"Emeraude," in French. Originally designed for a 65-hp Continental engine, Piel upgraded the design several times. The design even formed the basis for a production aircraft, the CAP Aviation CAP-10B.

Shankland had the plans for about two years before he started building. The Emeraude is not a kitplane—the builder constructs all the parts of the wood-and-fabric aircraft. Not only that, but the plans use the metric system and are written in French! Fortunately, many key terms had translations in Shankland's set (which isn't always the case with Emeraude plans). He installed a 150-hp Lycoming O-320, and the minimum avionics necessary for flight in the proximity of the Seattle Class B airspace. Construction took him almost four years, and he estimates he had about $35,000 in the aircraft. To get ready for the first flight of his aircraft, he bought a used Emeraude and flew it until his own was ready. He finally took "Esmeralda" (Fig. 7-17) for its first flight on July 4, 1999.

Ownership experience

Esmeralda resides in a hangar on a small private airport 6 miles from the Shankland home. Hangar rent is $180 a month. The airport itself is a bit on the primitive side, having no power or water available at the hangars, but the owners recently repaved the runway. No fuel is sold at the airport, but Shankland's Lycoming is

Fig. 7-17. *Rich Shankland and "Esmerala," his homebuilt Piel Emeraude.*

autofuel-compatible. "My Ford pickup has two 19-gallons factory-stock fuel tanks," he says. "I put a tee at the inlet of the stock mechanical fuel pump and ran a feed line up to a fender-mounted electric fuel pump. It has a one gallon per minute capacity and cost about a hundred bucks."

He carries basic liability insurance and ground-only hull coverage, which runs about $700 a year. He received the repairman certificate for his airplane, so Shankland performs his own annual inspections. "All but the specialized work on the engine," he explains. "I'll remove and replace the plugs myself, but I pay to have them cleaned and gapped." He found a local A&P to work with via a friend.

He estimates his typical annual maintenance costs run about $200. "I'm on my second set of tires, but the first ones wore out prematurely due to excessive camber," he says. "I'm about to reline my brakes for the first time, and the mags are going to need a routine check and tune-up." His major costs have come from his "boat anchor" transponder—"It's averaged over $100/year to keep it running. I bought it used and shouldn't have."

He flies about 60 hours per year. Shankland has taken Esmeralda to the Oshkosh fly-in twice, with detours taken to the East coast both times. On his second trip to Oshkosh, a wheel axle broke while taxiing on rough ground. The EAA Emergency Aircraft Repair Center helped get him repaired for the trip home, but he ended up redesigning and rebuilding the landing gear.

A summary of his ownership costs is shown in Fig. 7-18.

Advice

"If performance or cutting edge design is important, a homebuilt might fill the bill," says Rich Shankland. "Then there's those classics—you might find it a bit difficult to rent a 1935 Cessna Airmaster. So if that's what you have to have, you're gonna have to buy one."

Shankland feels that sole ownership of an aircraft is justified with over 125 hours a year utilization, but agrees there are other factors. "Pride of ownership or safety consideration can skew the numbers considerably."

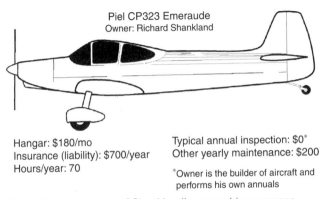

Piel CP323 Emeraude
Owner: Richard Shankland

Hangar: $180/mo
Insurance (liability): $700/year
Hours/year: 70

Typical annual inspection: $0*
Other yearly maintenance: $200

*Owner is the builder of aircraft and
performs his own annuals

Fig. 7-18. *A summary of Shankland's ownership expenses.*

8

Found It!

You've found the plane of your dreams—you hope. It's important to take a careful look at it before committing yourself. In this chapter, we'll examine the steps to take to find as much as possible about the condition and status of the aircraft and reduce your risks.

The process is fairly simple. First, you'll examine the airplane to see if there are any major problems with it, including looking through the logbooks and taking the plane for a test flight. Then you'll hire an A&P mechanic to perform a prepurchase inspection to get a good idea of its mechanical condition. Finally, you'll do some checking to ensure yourself that the seller indeed owns the airplane, and there aren't any encumbrances on it.

Most of the information in this chapter is aimed at those buying used aircraft—for instance, hiring a mechanic for a formal prepurchase inspection of a brand-new plane is a waste of money. Still, much of the advice applies, to some extent. Finding a flaw in a new airplane will be discouraging. Yes, it'll be covered by the warranty. But the dealer will fix the problem a whole lot faster if you refuse delivery (for example, giving them the check) unless the discrepancies are fixed.

EYEBALLING THE CANDIDATE

You probably don't know much about airplanes, other than how to fly them. But there are a few things you can do; some obvious signs that an airplane should be eliminated from consideration. Since a mechanic's prepurchase inspection is going to cost between several hundred and a thousand dollars, you'd like to pay for only *one*. Weed out the dross before shelling out the bucks.

So, how do you start? Just perform the most thorough preflight you've ever done in your life, fly the plane, and page through the logbooks to see how the plane has been maintained in the past. Through this process, make note of the questionable points, and make sure to mention them to the A&P before the prepurchase inspection.

First impressions

Start by standing 20 feet away and just looking at it (Fig. 8-1). Does the plane sit level? Is it maybe cocked over, leaning to the side like it's favoring one gear leg? For a fixed-gear Cessna, that's an immediate cause for worry. They've got one-piece steel landing gear. If a leg is bent, there's been one heck of a slam.

Other airplane types might lean merely due to a soft oleo strut. But why is it soft? Is it leaking? Make a note to check the logs. If they've been fiddling with that side's oleo, perhaps it isn't fixed yet. Then again, maybe one wing tank is full and the other empty. Check!

Walk around the plane and examine the way the cowling and fairings hang. Are they a little crooked? Might be some worn bolt holes or warped fiberglass. Look at the paint. Are some pieces painted differently? They might have been replaced from an aviation salvager. Nothing really wrong with that per se (the parts are usually quite airworthy), but you'll want to eventually know *why* replacement was necessary.

The paint

Move in and look more closely. How's the paint? Ratty and patchy?

Paint has two functions on an airplane. The first is decoration; it makes the plane look pretty. Second, it protects the underlying material—aluminum, wood, fabric—from the ravages of the environment.

We all prefer shiny new paint jobs. However, if cash is more important than cosmetics, you can find some pretty good deals on planes with cruddy paint. As long as the airframe is protected (Fig. 8-2).

True, you'll have to get an A&P's opinion on that. Poor paint on a fabric airplane can cause premature deterioration of the covering material. But the all-metal

Fig. 8-1. *Some problems with used aircraft are more apparent from a distance than up close. This Commander is clean, sitting level on its gear, and appears well-tended.*

Fig. 8-2. *Flaking paint doesn't make all-metal planes unairworthy. However, it should be studied closely to determine if the flaking is caused by corrosion.*

airplanes of the last 30 years do pretty good at standing up to the ravages of the weather. Nice paint doesn't make the plane fly any better. Still, if you've got a spouse who's grudgingly given approval to the aircraft purchase, buying something with grubby paint isn't likely to bolster their confidence. And shabby-looking planes are a bit embarrassing.

A spiffy paint job might hide rot underneath. Take a good look at the paint. Any funny curlicue ridges? Might be filliform corrosion. Is it flaking off anywhere; are the steel parts rusty? Might indicate prepainting preparation was poor or nonexistent. Or just plain hard use.

Exterior exam

Looking farther, try to get a feel for the material underneath the paint. Ripples/dents in an all-metal airplane? Red alert. Wrinkles in fabric? Warning flag. Examine the control surfaces, the flaps, exposed control cables, pulleys, even the bolt and screw heads. Look for rust and missing pieces. Move the surfaces around, checking for free play. Noises, too—a creak, scrape, or groan—might indicate a frozen bearing in a pulley or too-tight cables.

Look at any external, fixed control tabs. These are pieces of aluminum attached to the control surfaces to minimize control pressure in level flight. Many planes have them on the rudder, a number of them have tabs on the elevator, and a few carry a fixed tab on an aileron.

One bent at a steep angle indicates that the aircraft is misrigged or out of true. One on the aileron typically means a serious wing heaviness problem, as most airplanes have better means for correction. Check the wheels and tires. Look for the red stains of brake fluid—bad brakes will cost more to fix than worn tires. And you can change the tires yourself.

While you're crouching, check the belly. Most have some oil and soot stains, but excess of either is cause to wonder. If the paint looks a bit different on the belly of a retractable, there might have been a gear-up landing in the past. Not that the repairs would have just required a coat of paint—often the aluminum skin must be replaced, and that would show new paint.

Cabin and avionics

How's the cabin? Do the doors fit well? Are the seats well-upholstered and comfortable? Are they tight on their tracks? Any funny smells, like fuel or mildewed carpet? Does the panel have the avionics you want (Fig. 8-3)?

This last point is important. Some people figure, "Well, I'll buy a cheaper plane with obsolete avionics and upgrade the panel." Mention this to a bunch of aircraft owners, and watch for those who cringe. They've probably taken this route. It's not just the expense; it's the delay and frustration. The plane will be grounded for quite a while, and you'll take it back to the shop several times to get the little things straightened out.

The hassle is proportional to the complexity of the work. Replacing an older Nav-Com with a new model isn't all that bad, especially if you fly from an uncontrolled field where a radio isn't strictly required. But if you want a full-house up-to-date IFR suite, lean toward planes which are already equipped.

Two pieces of avionics to check on. First, see if the plane has an encoder-equipped (i.e., Mode C) transponder. Vast areas still don't require it, but the lack will limit

Fig. 8-3. *It's a lot easier to buy a plane with the avionics you want than to spend the hassle getting the panel upgraded.*

cross-country utility. Older planes like Cubs and Champs generally may not have them. If the plane was "certified without an engine-driven electrical system," a transponder isn't required within the 30-NM "veil" under and around Class B airspace. The plane can't actually enter Class B and C airspace, though.

The second piece of avionics is the Emergency Locator Transmitter (ELT). For all intents and purposes, all planes are required to carry one (the major exception is single-seat aircraft). Check when its battery is due for replacement.

How are the instruments? Are they readable? Is their glass cracked? What about the trim molding around the panel and the glare shield? Glance around and check the Plexiglas. Is it yellowed? Cracked? Crazed?

Work the controls. Do they feel sloppy? Too much play can mean loose or worn cables, pulleys, bearings, and bushings. Listen for squeaks and scrapes. Run the trim system up and down, looking toward the tail to verify operation.

The engine

The engine is an area that most of us will leave for the inspecting mechanic. But like the airframe, there are a few things you can do to weed out the also-rans. Ask the owner to remove as much of the cowling as practicable (Fig. 8-4). You can't judge the engine by peeking through the oil door.

Again, you're performing an extremely thorough preflight. Look for cracked hoses, frayed wires, gouged ignition cables, broken baffles, and leaks (Fig. 8-5). Look for anything that just doesn't feel right. Stains beneath cylinders. Rubbed or scraped spots. Strong fuel odor. Scraping or binding when the throttle or carb heat controls are worked.

Examine the exhaust system. Cracked pipes, mufflers, or heat exchangers yield expensive repairs. Look at the end of the exhaust pipe. Grayish or tan ash is normal for aircraft burning aviation fuel; it's a residue from the lead in the gasoline.

Fig. 8-4. *Some plane's cowlings open enough for an engine inspection, but others must be removed.*

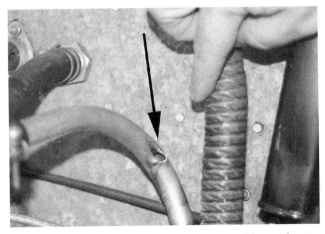

Fig. 8-5. *Hole in the breather tube? Yes, but this one is supposed to be there, in case the tube's end freezes up. Best to ask the mechanic during the prepurchase inspection, just be sure.*

Autogas leaves a black soot deposit. Oily or greasy deposits are something to ask the mechanic about.

Open the battery box. Any signs of stains around the fill caps? Perhaps the regulator isn't doing its job and the battery is boiling over. Any staining or deterioration around the inside of the box? Maybe the battery is leaking acid; bad news indeed if it's been dribbling on aluminum.

Don't forget the propeller. Any sharp nicks? Do both tips match in shape? If it's a constant speed model, are the blades loose or does the hub show an oil leak? If it's wood, is the varnish still shiny? Is the leading edge protection solid and whole? How's the spinner? Check for cracks and missing screws or bolts.

Test flight

Now comes the fun part: flying. Your main purpose in the test flight is to determine that everything works the way it should.

Have the owner start the engine. Watch if any weird little quirks are used. If the owner madly pumps the throttle during the start sequence, it might mean the carburetor accelerator pump isn't working right. Or it could mean the owner doesn't really understand the engine and might have caused excessive wear through ignorance. Leave the window or door open during start. Listen to the engine. Does it sound normal?

Perform a careful runup. Does it run fine on either magneto? Remember that properly set up engines will show a mag drop. Too often, a proud owner will say, "Look, only a 10-rpm drop on the left mag." In reality, the magneto timing is probably off. If no drop at all, the magneto switch could be faulty.

If the engine runs rough during the mag check or the RPM drops excessively, a number of things might be wrong. Express your regrets to the owner and taxi back to the ramp. You don't want to buy it with such a question mark. Heck, you don't even want to fly it. Let it get fixed out of the current owner's pocket, not yours.

The owner may not let you perform the takeoff or landing. Who can blame them? They know absolutely nothing about your skill level. They don't know how you'd handle a takeoff emergency, or even if you have enough hours to keep the plane under control. When I sold my 150, I turned the controls over to one prospective buyer when we reached pattern altitude. He was unable to even fly straight and level. I suspect he'd seen my for-sale ad and figured it was a cheap way to take his first airplane ride!

Letting the owner do the work lets you monitor the aircraft's performance. Watch the engine gauges during takeoff. Does the engine come up smoothly to full power? For a fixed-pitch propeller, the RPM should be around 80 percent of its redline at the beginning of the takeoff roll. From that point, it should increase gradually as airspeed rises.

Get a feel for the rest of the aircraft, too. Does it seem to be weaving around a bit? Check the crosswind, but maybe the gear is a little out of alignment. The usual cause is a hard landing. Be alert for any shaking up front on a trigear airplane. One of their banes is a tendency for nosewheel shimmy. Once the plane breaks ground, check the rate of climb. Considering the density altitude, does it seem to match the book? Remember, though, that the "book values" were determined by expert test pilots on brand-new airplanes. However, if published figures claim an 800 fpm rate of climb and the gauge only shows 400, there's cause for question.

The owner should let you take over at altitude. You don't have to perform the whole private pilot test repertoire; just do turns at various roll rates and banks (Fig. 8-6). How does the airplane feel? Are the controls tight and smooth, or sloppy and lifeless?

Set cruise power, trim up the elevator and release pressure on the controls, including the rudder. Does the airplane track straight? Most planes will eventually

Fig. 8-6. *The proof is in the flying. A test flight can tell you a lot about the condition of the plane's systems.*

drop a wing slightly and start a gradual turn. Especially watch the slip-skid indicator. Misrigging in yaw can be irritating and cause misbehavior during stalls.

Verify everything works, especially the avionics. Check the VOR against a local station. See if the GPS gives your proper position. Fly an ILS approach if the plane is so equipped. Tune the radio. Are both other aircraft and FAA facilities coming in loud and clear? Cycle the flaps/gear/propeller, monitoring for proper operation and rude noises. Above all, monitor yourself. How do you feel in this plane? Are you happy with the way it flies? Could you spend a week flying cross-country without feeling the urge to take a fire-ax to the interior?

Back to the airport. Still looking good? Time to look at the logbooks.

Checking the records

Logbooks are fascinating things.

I used to take my 150's books out once in a while and just scan through them. You can see where the plane's been; what's been done with it. You can touch the signatures of the previous owners and wonder what they're flying now. Just from the number of entries and their thoroughness, you can tell who took care of the plane and who didn't.

Examining the logs of an unfamiliar airplane is important to get a feel for the condition of the aircraft. But the logs don't tell you right out. It takes a bit of detective work and inference to extract the true background of the airplane. For instance when I bought my 150, the sign in the window said, "Never a Trainer." I was skeptical. I didn't believe that a 20-year-old Cessna 150 could have escaped being flogged around the pattern by students at some point in its life.

Sure, the use of the aircraft isn't indicated in the logs. But there were entries about annuals being performed by "XYZ Flying Club." There were periods when the plane flew up to 300 hours between annuals. That wasn't just a happy owner flying around on the weekends. In addition to such routine entries, the FAA requires that any repairs or inspection also be noted in the logbook. However, the FAA doesn't require that the *cause* of the repair be mentioned. It's in your court to read between the lines.

This entry in my 150's logs set flags waving: "Replaced leading edge skin between outboard ribs." Why would they do that? Damage, obviously. And damage on the leading edge generally means the airplane hit something. Probably on the ground, since midairs generally negate the need for further log entries.

Was the leading edge damage significant? I didn't think so. It had happened 15 years earlier. Had it been more recent, I would have worried about damage that the repair mechanic might have missed. Hitting a pole with the wingtip can cause more than external damage; the leverage can tear metal in the wing's root and even the fuselage. In my case, the plane had undergone 15 annual inspections since the repair. I figured one of the IAs would have found any additional damage.

Don't forget the engine and propeller logs, as well. They can tell their own stories of problems and damage. If you find an entry in the engine log regarding checking the "crankshaft runout," beware. This is typically done after a propeller strike, the ramifications of which were discussed in Chap. 2. It's used to determine if the shaft was bent by the propeller impact.

By the way, not all aircraft have a separate log for the propeller. For a long time, planes with fixed-pitch props kept a combined "engine and propeller" book. The FAA now encourages separate logs, even for fixed-pitch props.

Engine cylinder compression is checked at each annual; the results are required to be entered in the engine log. The test procedure isn't at all like that used for cars. You'll see a set of numbers such as 80/72 (or they may be listed the opposite way, 72/80) listed for each cylinder. We'll cover what these numbers mean in a Chap. 13. For right now, you should know that if the smaller number is below 60, cylinder, ring, or valve work is probably required.

Track the compression values over a couple of annuals. If one cylinder's compression drops suddenly, beware. There are a few legitimate, noncritical reasons for a temporary reduction. But you might end up buying a new cylinder all too quick. Every replaced aircraft component should have been accompanied by paperwork to prove that it was fully FAA-approved and airworthy. If the engine log shows that a magneto was replaced, the "yellow tag" that came with the magneto should have been retained. Look for these tags. Don't forget to check on the ADs for the airframe, engine, propeller, and appliances, too.

Sometimes putting all three logbooks together can be enlightening. Say you're looking at a 1978 Mooney. You notice that the propeller log starts in May, 1990 with the installation of a new propeller. Why was it replaced? The engine log shows the engine was torn down and inspected at about the same time. And the airframe log shows the belly skin was replaced as well. Shouldn't take a lead pipe to convince you that the plane was landed gear-up in early 1990.

Does it make a difference? That far back, probably not. But if the owner is claiming "No Damage History," you've got leverage when you start dickering over price. Don't immediately suspect the owner of trying to pull a fast one. He may not have owned it at the time and may not have checked the logs carefully when he bought the plane.

Should you reject planes, which have suffered major damage in the past? No, but you must assure yourself that the aircraft was properly repaired. Take the Piper Warrior in Fig. 8-7—it flew through a set of power lines on final approach at night, landing safely afterward. Note that the wing leading edges have major wrinkles, and the vertical tail is knocked back almost 90 degrees. But the plane is repaired and still operational. The current owners knew about the history, and emphatically state that the plane flies just fine.

Again, when checking the logs, you don't have to be all-knowledgeable about aircraft repair. Just pick out sections that you find unusual, and talk to an A&P. One nice feature of the Internet is that the National Transportation Safety Board (NTSB) maintains its accident reports online—you can search for the N-number of the plane you're looking at. If it's been involved in anything, but the most minor accidents, the report should be listed. Go to *www.ntsb.gov* and click the "Aviation" tab.

The logbooks will also tell you of the modifications that have been performed on the aircraft. If the owner tells you the plane is approved for autogas, you should be able to find an entry recording execution of the appropriate STC. There should

Fig. 8-7. *This Piper Warrior flew through a set of high-tension lines on final approach. It was subsequently repaired and apparently flies well, demonstrating that even major damage can be repaired successfully.*

also be a copy of the FAA Form 337 that was executed and sent to the nearest Flight Standards District Office (FSDO). All modifications require the filing of a Form 337; if the aircraft has been modified but the paperwork hasn't been filed, the aircraft isn't legally airworthy. If you bought such a plane, you'd have to have a mechanic check the mods and file the appropriate paperwork.

Should you wish to verify that the paperwork has been filed and accepted by the FAA, you can request a copy of that aircraft's records. Go to *www.faa.gov*, find the search window, and enter, "copies of aircraft records." The FAA will send you a CD-ROM with copies of all FAA paperwork on the airplane for $6.25. If you prefer hard copy, the basic search costs $2, the first page costs an additional $0.25, and each additional page is just a nickel more. For more information, call the Aircraft Registration Branch at 405-954-3116, or toll free at (866) 762-9434.

Finally, as you go through the logbooks, check on when the aircraft last had an annual inspection. If the annual is due next month, that's at least another $500 (and probably more) that'll be coming out of your pocket almost immediately. There are other required inspections as well; the transponder system must be checked every 24 months, and, for IFR operations, a static system test is required at the same interval.

Engine matters

As I've mentioned earlier, the condition of the engine is one of the major price drivers, especially for smaller, simpler planes. If your prospective purchase has a "high time" engine, it's good to be cautious. Expect to pay less.

Just remember, if the manufacturer lists an 1800 hour TBO and the engine has 1700 hours Since Major Overhaul (SMOH), that doesn't mean you'll get only a hundred hours of flying out of it. I've a friend who rebuilt a Piper Tripacer. The engine was 400 hours past the TBO. Yet he didn't rebuild it. The Lycoming runs strong and has shown no signs of distress. The rebuilt airplane has been running happily on that "run-out" engine for several years now.

How an engine has been operated is more important to its longevity than how long. The Lycoming in a 152 is going to perform a lot of touch-and-goes; a bunch of full-power climb outs immediately followed by idling glides. It's under a lot more stress than one used in a fish-spotter, who spend almost all their time at cruise RPM. They've been known to go as far as a thousand hours past the recommended TBOs.

Still, TBO-busting is a nerve-wracking process. A good number of perfectly-good engines have been overhauled just because the owner got jittery as the tachometer's hourmeter neared the magic number.hash But even if the engine in your prospective purchase has only a few hours since its last overhaul, you aren't home free. The quality of overhauls varies drastically.

Two local guys bought a Cessna 140. Its Continental C-90 had 300 hours since overhaul; that's considered a low-time engine. They spent $3000 on the engine. Some of the internal parts were trash. And they didn't get that way in 300 hours, either.

Any licensed A&P can overhaul an engine. How good the overhaul is depends on how finicky—and to be brutal, honest—that mechanic is. The procedure seems simple enough. The overhaul manual for the engine specifies the maximum amount of wear a part can show and still be used on the engine.

Yet there are ways around that; ways to stretch the specs enough to get marginal parts to pass. It may be done as a favor to a friend, to give his engine an "overhaul" at minimum cost. It may be done to maximize profit. The person performing the overhaul may just have been lazy or sloppy.

So if an engine is listed as being recently overhauled, dig a little bit and try to determine how good an overhaul. If it was performed by the factory or one of the nationally known engine shops, so much the better. But local mechanics can perform as good of a job. Be prepared to ask your prepurchase inspection A&P about the overhauler. Ask the owner for the receipts from the overhaul that will help determine what work was done.

There's a way to ensure that the overhaul is done to your standards. Buy a plane with a broken or run-out (i.e., past its TBO) engine, and have the engine rebuilt at a quality shop. It'll require a greater initial cash outlay, but you'll end up with an engine you can trust.

THE PREPURCHASE INSPECTION

Looking the plane over yourself is a great start, but with the kind of money at stake, a prepurchase inspection is a very, very good idea. Remember the case study in Chap. 6—Mike Furlong bought a "Ramp-Queen" Stinson after a talk with the owner and a cursory look-over. It took him nine years and almost six times the original purchase price before he flew his new plane.

Sometimes people get lucky. When I bought my 150, I hadn't flown in seven years. I'd never flown in the state, in which I was then living; I had no contacts nor knew where to find an A&P. I read the logs, gritted my teeth, and handed over a check. It turned out OK. But in retrospect, it wasn't the best way to go.

How do you find a mechanic to do the prepurchase inspection? Start by looking in the aircraft logbooks to find who did the last annual. Scratch that mechanic off your list. He or she is probably competent. But if they missed something at the last annual, they just might miss it again. Or if they find it, they may be too embarrassed to admit it.

If you have any friends who own airplanes, see if they'll recommend someone. Otherwise, just call some local FBOs. All shops are familiar with prepurchase inspections and should be able to give you quotes over the phone. If the plane is an unusual one, spend some time to locate a mechanic with experience in that model (Fig. 8-8).

Depending upon the complexity of the aircraft, a prepurchase inspection will probably run from $250 to $750. At the utmost minimum, you want the mechanic to give the engine a clean bill of health (Fig. 8-9), plus determine if the plane is legal; that is, the ADs are up to date, all maintenance was performed and logged properly, no life-limited parts (some hoses, ELT batteries, and the like) have expired, all required markings and placards are in place, all inspections (annual, static system, transponder, and manufacturer-required accessory checks) are current, and so on.

That level of inspection will make sure you (probably) won't have to deadstick on the way home, and ensure any flying you do is legal. There's a lot more that *should* be done, though. If your mechanic is experienced in the aircraft model, he or she should know where the potential problem areas lie. Where corrosion tends to appear. What tends to form cracks, what tends to break, and what tends to wear?

Fig. 8-8. *In considering a rare airplane like this Varga Kachina, it can be difficult finding a mechanic familiar with the breed for the prepurchase inspection.*

Fig. 8-9. *A compression check is a standard test during a prepurchase inspection. For more details, see Chap. 13.*

There are few "trivial" problems on an airplane. Broken window latch? Seems minor, except new ones might only be available through the factory. That simple latch might cost $250 or more. Plus labor to install it. You want to know as much as possible about the airplane's condition before you're the one responsible for it. The prepurchase inspection is not the place to scrimp. The best ones are where the mechanic performs the equivalent of an annual inspection. Except, of course, he or she won't do any cleaning, lubricating, or repairing.

If the aircraft's annual is due, you might be able to work a deal with the seller: You'll pay the inspection cost of the annual, if the seller pays for the repair portions of it.

Prepurchase inspections aren't covered by any regulation, hence the results aren't noted in the aircraft logbooks. Instead, the mechanic should give you a summary of his assessment of the aircraft's condition. The mechanic should be able to give you an informal estimate as to the costs to repair based on the deficiencies found.

One thing to keep in mind: Prepurchase inspections are not magic. There are a lot of conditions they won't turn up, and can't discover unless the plane is completely disassembled. No matter how honest the owner or how good the inspecting A&P is, you might still end up with expensive repairs soon after purchase.

WHO OWNS IT?

Before you contemplate handing a large check over to this aircraft-selling stranger, you'd better ensure that he or she owns the airplane in question! Maybe the owner just didn't bother to file the required paperwork when the airplane was purchased and the title is thence in question. Or, perhaps the plane is collateral in a bank loan. Maybe it's even stolen.

If you buy the plane and it turns out that some third party actually had valid title on the plane, they get it. No ifs, ands, or buts. You're out the purchase cost, unless you can find the guy you bought it from and extract it from him. Good luck.

The title search

Like a prepurchase inspection, verification that the seller has legal authority to sell you the airplane can save you a bundle later. You can find this out through a title search.

Unlike a lot of financial/legal issues, understanding title searches is simplicity itself. The search just tells you who the current legal owner of the aircraft is. If the name differs from that of the snake-oil salesman with his hand out for your check, *run*.

Unlike a prepurchase inspection, title searches are low-cost and painless. The price is generally less than $50; the results are usually received in a day. Get a title search, no matter who's selling the aircraft—family member, friend, FBO, broker, or national sales firm. If the seller tries to talk you out of a search, run. The penalty for error is just too severe.

Get one. No exceptions.

Title search companies

Many aircraft publications include advertisements for companies that perform title searches. In Trade-A-Plane, for instance, look under "Aircraft Title Service/Legal Services." AC 8050-55, "Title Search Companies," is available from the FAA. It lists those companies approved to perform title and record searches. Or you can just enter "aircraft title search" on your favorite Internet search page.

The AOPA also performs this service, and they also provide many other useful prepurchase services. Their number is 1-800-USA-AOPA (872-2672), or go to *www.aopa.org*.

Paying off the seller's bank

If the owner used the aircraft as collateral for a loan, the finance company will be listed as the owner of record. Buying the plane in this case will be similar to buying a used car where the seller still owes money. You'll have to actually pay the bank, who'll sign over the title and pass the cash left over from paying off the loan to the seller.

THE DEAL

Did it pass the tests? If so, we're on the home stretch.

Dickering

Ever bought a used car? The negotiation process for a used airplane works similarly. The owner claims the plane is a rarity, in superb condition, and you'll argue that it's a piece of junk that you might be willing to take off their hands.

Actually, it's best not to go *quite* that far, at least with aircraft. If the plane meets your requirements, you don't want to antagonize the owner. Small egos are rare in aviation. Tick off the owner, and he may decide not to sell to you at any price.

And what *should* that price be? You've already done your research and know what equivalent aircraft are selling for. Your mechanic has identified several items that'll need repair, so deduct the cost from the estimated worth.

Base your dickering with the owner on the aircraft's estimated worth. It goes without saying that you'll want to offer less; there's nothing worse than making an offer that the seller accepts with lip-smacking relish. Have your research materials available. Be ready to show the owner some advertisements for equivalent airplanes; point out the flaws your mechanic found. Cold-blooded as it may sound, your best hope for saving money lies in an owner who didn't do any research on what his plane is worth.

Which makes dealing with brokers or FBOs that much less satisfying. They'll not only try hold out for top dollar, their commissions and markups run the price even higher. Overpaying for anything, including airplanes, leaves a bad taste in one's mouth. You'll make a lot better deal with a private seller. With private individual or aircraft broker, you'll sometimes come to an impasse. The seller has come down as far as he or she will, and it's thousands above what you want to pay. Your response can take a number of forms.

The "book" response here, of course, is to walk away. You know what the airplane's worth and shouldn't pay a penny more. Unfortunately, in the real world, things aren't that easy. Sometimes you really want that particular airplane. Maybe it's a rare model you've looked high and low for. Maybe it's equipped exactly the way you want. Should you bite the bullet and pay the asking price?

It's impossible to make that decision in advance. Hopefully, you know the maximum you can afford. And no matter how good the airplane is, it'll probably cost more to own than you expect. Don't put yourself even further in the hole by paying more than is comfortable.

There are a couple of alternatives. First, you can politely take your leave, giving your name and phone number to the owner. If a month or so goes by without the plane selling, they may be more interested in your offer. Second, you can offer to pay the asking price, but make the deal contingent on associated issues. For instance, you might suggest the owner pay for a fresh annual inspection. Even if the plane had been inspected just six months earlier, a new annual would push the date of the first one *you* pay for that much farther back. If the plane's being sold through an FBO, they can often get the annuals at a discount rate.

Other issues are possible with a private owner. Perhaps you're three years down on the hangar waiting list, and the seller already has a hangar at your home field. Offer to pay the higher price if he or she subleases the hangar to you for the next two years. Or maybe the seller of a rare plane has a stock of spare parts; make those part of the deal if you can't get the selling price down.

The handover

Hopefully, eventually, you'll agree on price. The next thing to work out is how to pay.

Would you accept someone's personal check for $40,000? Especially when they're going to take possession of valuable (and highly mobile) property?

I bought my 150 with a certified check; the cost is minimal and the seller's risk low. When I sold the plane, the buyer paid cash! A veritable pile of $100 bills. I hadn't known he was going to do that. We'd agreed for the handover at 4:30 p.m. on a Friday. At 4:40 p.m., I had a bulging envelope full of (freshly counted) bills and a sinking feeling that I'd be staying up all weekend standing guard on it. I get nervous when I have a $100 bill in my wallet. Fortunately, I made it to a nearby bank before they closed.

Not all purchases are this simple. Sometimes, the seller wants to see a little "earnest money" up from. Or the seller promises to have the aircraft annualed if you agree to buy it. Or the seller lives in one location, you live in another, and the bank that holds the lien on his airplane is located far away from you both. Obviously, handling the paperwork and money transfers in these cases is awkward. An escrow agency can come to the rescue. They act as a disinterested third party, ensuring that the money and title are not transferred unless agreed-on conditions are met and that all required documentation is properly submitted.

If your transaction is simple and you tackle it yourself, make sure the plane has the following before you hand over the lucre:

1. Airworthiness Certificate (FAA Form 8100-2)
2. Weight and balance/equipment list
3. Flight manual or operating limitation
4. Airframe log
5. Engine log
6. Propeller log (or combined engine-propeller log)

Transfer of the aircraft is handled using FAA's Aircraft Bill of Sale, AC Form 8050-2 (Fig. 8-10). This two-part form is available at FAA offices and many FBOs. Otherwise, you can download it, do an internet search for "8050-2," and print out two copies.

When everything with the aircraft is to your satisfaction, give the owner the money and have him or her fill out and sign the Aircraft Bill of Sale (two copies, if

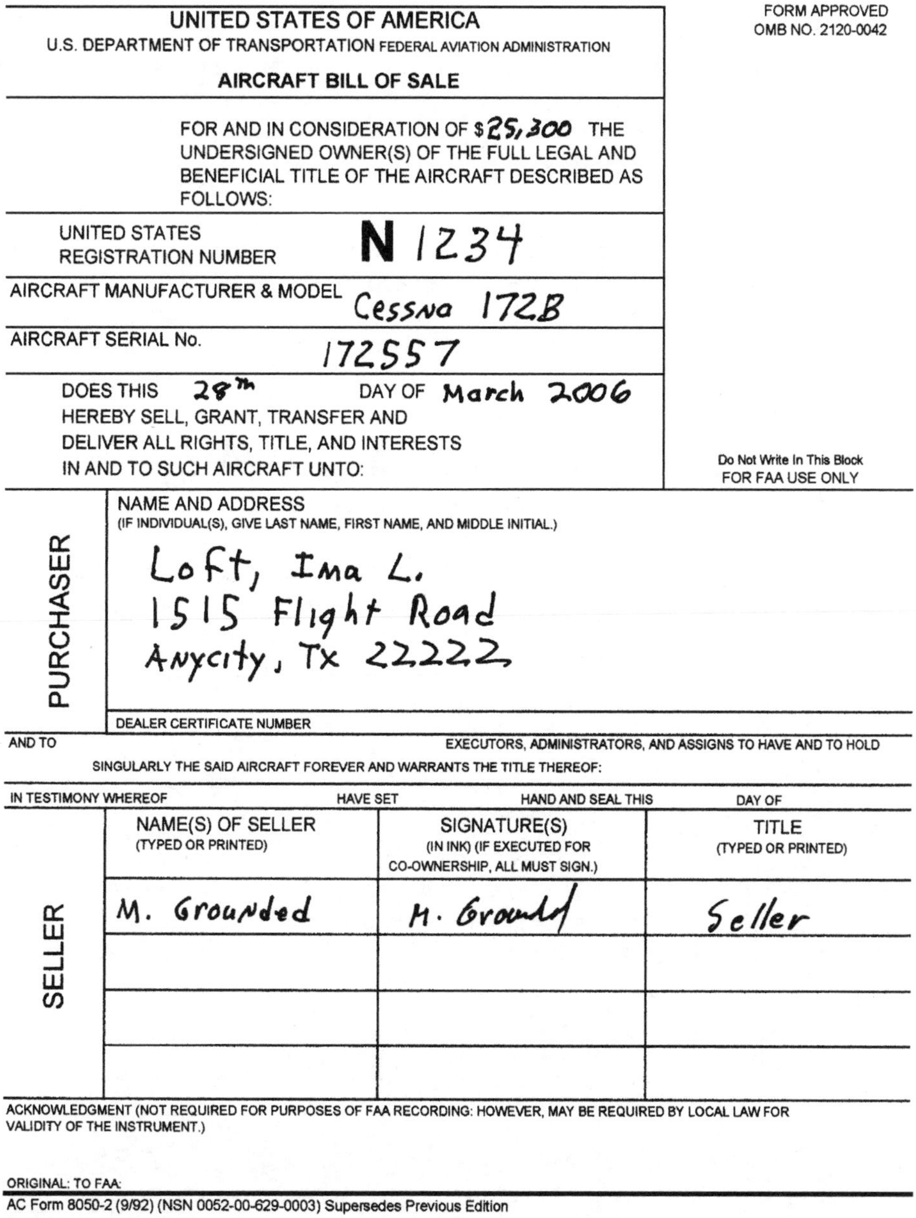

UNITED STATES OF AMERICA
U.S. DEPARTMENT OF TRANSPORTATION FEDERAL AVIATION ADMINISTRATION

AIRCRAFT BILL OF SALE

FOR AND IN CONSIDERATION OF $**25,300** THE UNDERSIGNED OWNER(S) OF THE FULL LEGAL AND BENEFICIAL TITLE OF THE AIRCRAFT DESCRIBED AS FOLLOWS:

UNITED STATES REGISTRATION NUMBER	**N 1234**

AIRCRAFT MANUFACTURER & MODEL	**Cessna 172B**

AIRCRAFT SERIAL No.	**172557**

DOES THIS **28ᵀᴴ** DAY OF **March 2006** HEREBY SELL, GRANT, TRANSFER AND DELIVER ALL RIGHTS, TITLE, AND INTERESTS IN AND TO SUCH AIRCRAFT UNTO:

Do Not Write In This Block
FOR FAA USE ONLY

PURCHASER

NAME AND ADDRESS
(IF INDIVIDUAL(S), GIVE LAST NAME, FIRST NAME, AND MIDDLE INITIAL.)

Loft, Ima L.
1515 Flight Road
Anycity, Tx 22222

DEALER CERTIFICATE NUMBER

AND TO EXECUTORS, ADMINISTRATORS, AND ASSIGNS TO HAVE AND TO HOLD

SINGULARLY THE SAID AIRCRAFT FOREVER AND WARRANTS THE TITLE THEREOF:

IN TESTIMONY WHEREOF   HAVE SET   HAND AND SEAL THIS   DAY OF

SELLER

NAME(S) OF SELLER (TYPED OR PRINTED)	SIGNATURE(S) (IN INK) (IF EXECUTED FOR CO-OWNERSHIP, ALL MUST SIGN.)	TITLE (TYPED OR PRINTED)
M. Grounded	M. Grounded	Seller

ACKNOWLEDGMENT (NOT REQUIRED FOR PURPOSES OF FAA RECORDING: HOWEVER, MAY BE REQUIRED BY LOCAL LAW FOR VALIDITY OF THE INSTRUMENT.)

ORIGINAL: TO FAA:

AC Form 8050-2 (9/92) (NSN 0052-00-629-0003) Supersedes Previous Edition

Fig. 8-10. *Bill of Sale form.*

you're using a downloaded form). Make sure the seller prints his or her name in the "Name(s) of Seller" block—the FAA has gotten picky about this, lately.

Congratulations. You have an airplane!

Now comes the hard part: actual ownership. The rest of the book covers the legal, mechanical, and emotional aspects of your new possession.

Other sources

Buying an airplane is a nerve-wracking, complex operation. You won't hurt my feelings a bit if you decide to read what other authors have to say about the process. Here are some other McGraw-Hill books, which cover aircraft buying:

Aviation Consumer Used Aircraft Guide, catalog #2441.

The Illustrated Buyer's Guide to Used Airplanes, Catalog #2462, by Bill Clarke.

As mentioned before, there are also books specializing in particular aircraft or aircraft lines. In addition, a type club might be able to provide helpful hints. If you're an AOPA member, the organization has a nifty little packet with tons of information and advice, plus copies of all the required FAA and FCC forms. Call (800) 872-2672 and select the member services department.

CASE STUDY: A HEAVY-HAULING PARTNERSHIP

Mike and Cindi's Bentz's choice for their first airplane was fairly simple: With three kids, they needed something more than the ordinary Cessna 172 or Piper Archer. With a vacation cabin to be built on an island in Washington State, they needed a plane that could carry everything for the new home, *including* the kitchen sink.

Their solution: a 1967 Piper Cherokee Six, a seven-seat 300-hp behemoth with a near 1500-pound useful load (Fig. 8-11). "When we get out of the plane," says Mike, "It looks like a clown car." The airplane is fully IFR equipped, with two Terra comm radios, separate nav receivers, ILS, ADF, autopilot, a VFR GPS, and a host of other gear, including a STOL kit.

Their combined salaries (Mike is an aerospace engineer, Cindi is the director of a preschool) weren't enough to allow them to comfortably support the airplane themselves. After a failed try as a leaseback, Cherokee N4082R found a happy niche as the centerpiece of a partnership.

Fig. 8-11. *Mike and Cindi Bentz originally bought this Cherokee Six for leaseback, but eventually sold partnership shares.*

Acquisition

Bentz's purchase of the big Piper has all the makings of a classic horror story. "I was pretty naïve when I bought it," says Mike Bentz.

But he did just about everything right. He'd found the Cherokee in Oklahoma via a search of Internet ads. He arranged a lease-back setup with a local flying club. He arranged a loan through his credit union. An Oklahoma A&P he'd hired to examine the airplane—which had a current annual—gave it a thumbs-up. He was ready to buy.

First problem: At the very last minute, the financial institution decided that the planned leaseback put the airplane into a commercial situation—and they denied the loan as being too risky. Fortunately, the credit union quickly substituted a home-equity loan instead, and Mike Bentz was on his way to Oklahoma. He paid the $65,000 they'd agreed upon, climbed into the airplane, and started back to Seattle. But it was winter, a tough time to fly back over the Rockies, especially in an unfamiliar airplane. He was forced to abandon it for a while in Sacramento, returning later to bring it home.

Before they put the Piper on their rental line, the flying club required to have their own A&Ps take a look at it. "The mechanics just shook their heads," says Bentz with a grimace. The inspection revealed some unpleasant surprises. The alternator belt came from an auto-parts store. Corrosion was evident. Airworthiness Directives were logged as having been complied with, but the faulty parts were still on the plane. A landing light connector had been replaced with a bulb socket and the base of a bulb with the glass broken away and wires soldered in place. All four fuel tanks vents had been sawed off flush with the wing skin and covered with screws. The previous annual was not only a joke, but was improperly logged and thus never had been valid.

Still, all it took was money. *Fifteen thousand dollars* later, Cherokee 82 Romeo was approved to sit on the rental line. And that's about all it did: Sit. The (frightfully expensive) insurance required a high-performance endorsement, and few people needed the massive load-carrying capacity the Cherokee Six provided.

"We knew we had to do something," remembers Mike. So they took the Cherokee off the leaseback line and formed "Mid Life Crisis Aviation," a Limited Liability Company (LLC), to operate the plane as a partnership.

Partners

One of the first persons to buy in was one of the members of the flying club. "I was prepared to start flying it when Mike took it off leaseback," says Randy Smith. He owned property on the same island as the Bentz family, and needed a load-hauling airplane for exactly the same reason. "I had actually looked into several other airplanes," he says, "but liked the Six because of its large CG range and gross weight capability."

"I've been in two partnerships, and this one is the best," says Smith. He had been previously in a three-way partnership on a Cherokee 140. "Two of us were much more involved than the third person, and it made things difficult."

The other two memberships in the club have changed hands over the years. Bob Carter is the latest owner of one of them. "I joined for two main factors," says Carter, "Capability and availability." The capability aspect was driven home when he witnessed the previous owner returning from a camping trip. "He had a whole campground in the back of the airplane" (Fig. 8-12). One of Carter's favorite uses is taking out-of-town visitors to see the sights of Western Washington state. He flies to the San Juan islands, north of Seattle on occasion, but most visitors want to see the main tourist attraction: Mount St. Helens. Either trip would take three or more hours by car, but the big Piper takes only 45 minutes.

Recently, the owner of the fourth partnership put his portion up for sale. It was on the market for months. They got plenty of nibbles, but the insurance requirements (500 hours total time, or an IFR rating, or an extensive checkout) eliminated most of them. Eventually, Mike Bentz bought him out. The portion was put in Cindi Bentz's name to allow her to act as an officer of the LLC. The Bentzes may continue to try to sell the quarter-ownership, or the remaining members of the partnership may buy the quarter together and convert the partnership to a three-way (Fig. 8-13).

Ownership experience

As an older airplane (~4000 hours total time), Cherokee 82R has had its share of maintenance problems. "We went through three starters one year," remembers Mike Bentz. His family was stranded on the island on one of the instances. Several other items broke around the same time, and one of the previous partners got tired of the hassle and dropped out.

A number of ADs affect the airplane, many of them are of the "Inspect every 100 hours" variety. One requires an eddy-current inspection of the constant-speed prop at every annual, and eventual replacement of the prop. A bad oil-temperature gauge required $1800 in troubleshooting labor and parts.

Fig. 8-12. *Not only does the big Piper carry a heavy load, it has a huge baggage door to help loading and unloading.*

Fig. 8-13. *"Mid Life Crisis Aviation" is the company that runs the partnership on the Cherokee. From left to right, the partners are: Randy Smith, Bob Carter, Cindi Bentz, and Mike Bentz.*

Annual inspections have revealed items like bad magnetos, muffler problems, a failed vacuum pump, and a dead transponder. A typical annual inspection runs about $4000, but have been as high as $8000 since the partnership started.

The same A&Ps as when the plane was a leaseback usually are hired for the annuals. The partners tried an owner-assisted annual one year. Unfortunately, it was a year where quite a few things went bad, and the partners ended up driving out to a remote airfield for a month. Randy Smith says, "Flying out to another airport for an owner-assisted annual was a bad idea."

The three current partners all live in close proximity to the airport, "One on downwind, one on crosswind, and one on final," jokes Mike Bentz. The Cherokee is kept at an outside tiedown costing $85 a month. As is typical for older airplanes, the seals leaked a bit of water during Seattle-area winters, but a custom-fitted cockpit cover takes care of it. The members pay monthly dues of $100, plus a fee of $50 (dry) per flight hour. The $50 includes a $15 allocation toward engine overhaul. The dues and per-hour rate cover the fixed costs and much of the annual cost, but occasional levies of additional money must sometimes take place.

Utilization of the airplane average 125 hours per year. While the partnership bylaws allow a rotating schedule to equalize access, the utilization rate is low enough where this hasn't yet been invoked. The annual insurance premium for coverage of four partners runs about $2350.

Advice

The key point: The plane does what its owners need it to do. Most FBOs balk at the prospect of renters flying into rough grass airports. But two of the owners take the STOL-kitted Cherokee into the rough grass airport on the island with their vacation properties, carrying up to seven people or a half-ton of construction supplies at over 150 mph. A trip that would literally take all day by car and private boat is reduced to a half-hour jaunt.

The partnership members recommend the scheduling services offered by Aircraft-Clubs.Com (*http://www.aircraftclubs.com/*). For just $60 a year, it allows the partners to schedule the plane to match their needs.

Mike Bentz recommends that one member handles the maintenance issues, as it prevents mechanics from getting caught in a tug-of-war between members. "I couldn't afford a plane without a club or a partnership," says Randy Smith, "and I prefer a partnership." "It's nice to get comfortable with one airplane," adds Mike Bentz. Says Cindi Bentz, "It's been awesome and fun."

A summary of the ownership of N4082R is presented as Fig. 8-14.

1967 Cherokee Six
Owner: Midlife Crisis Aviation
Situation: Four-owner partnership

Tiedown: $85/mo
Insurance: $2350/year
Hours/year: 125
Member dues: $100/mo

Inspections: $1500
Other yearly $2500–$6000
maintenance:

Fig. 8-14. *A summary of Midlife Crisis Aviation's costs of ownership.*

9

The First Week

Now you've done it.

COMPLETING THE PAPERWORK

There are two government entities you'll have to placate: The FAA and your state's aircraft department. Depending upon how you're going to use your plane, you may have to file paperwork with the Federal Communications Commission (FCC) as well.

The FAA

You got the previous owner to fill out two copies of the Aircraft Bill of Sale form (or used the formal two-part form). The FAA gets one copy and you get to keep the other. Don't lose your copy; other than the obvious legal issue, you might need to wave it at some local or state agencies. More on that in a bit.

The other bit of FAA paperwork is AC Form 8050-1, Aircraft Registration Application (Fig. 9-1). This form is not available for downloading at the present time, and the FAA doesn't allow photocopies. You need to obtain it directly from the FAA, from either the Aircraft Registration Branch in Oklahoma City or from your local FSDO.

The seller should have kept the old registration since he or she will need it to notify the FAA that they've sold the plane. The pink copy included with Form 8050-1 is a temporary registration; immediately place that in the aircraft. You're covered for the 90 days or so it'll take for the FAA to send you a permanent registration.

The pink temporary registration has "OPERATIONS OUTSIDE THE UNITED STATES ARE PROHIBITED BY LAW" printed on it. Simple enough—don't fly the airplane to another country, such as Canada or Mexico, until the permanent registration is on-hand.

The FAA stocks copies of AC 8050-1. Since the form is also used for address changes, pick up an extra or two. To formalize the transfer of ownership, send the

FORM APPROVED
OMB No. 2120-0042

UNITED STATES OF AMERICA DEPARTMENT OF TRANSPORTATION
FEDERAL AVIATION ADMINISTRATION-MIKE MONRONEY AERONAUTICAL CENTER
AIRCRAFT REGISTRATION APPLICATION

CERT. ISSUE DATE

UNITED STATES REGISTRATION NUMBER N 1234

AIRCRAFT MANUFACTURER & MODEL
Cessna 172B

AIRCRAFT SERIAL No.
17255557

FOR FAA USE ONLY

TYPE OF REGISTRATION (Check one box)

☑ 1. Individual ☐ 2. Partnership ☐ 3. Corporation ☐ 4. Co-owner ☐ 5. Gov't. ☐ 8. Non-Citizen Corporation

NAME OF APPLICANT (Person(s) shown on evidence of ownership. If individual, give last name, first name, and middle initial.)

Loft, Ima L.

TELEPHONE NUMBER: (101) 555-7777

ADDRESS (Permanent mailing address for first applicant listed.)

Number and street: 1515 Flight Road

Rural Route: P.O. Box:

CITY	STATE	ZIP CODE
Anycity	TX	22222

☐ **CHECK HERE IF YOU ARE ONLY REPORTING A CHANGE OF ADDRESS**

ATTENTION! Read the following statement before signing this application.
This portion MUST be completed.

A false or dishonest answer to any question in this application may be grounds for punishment by fine and / or imprisonment (U.S. Code, Title 18, Sec. 1001).

CERTIFICATION

I/WE CERTIFY:

(1) That the above aircraft is owned by the undersigned applicant, who is a citizen (including corporations) of the United States.

(For voting trust, give name of trustee: ____________), or:

CHECK ONE AS APPROPRIATE:

a. ☐ A resident alien, with alien registration (Form 1-151 or Form 1-551) No. ________

b. ☐ A non-citizen corporation organized and doing business under the laws of (state) ________ and said aircraft is based and primarily used in the United States. Records or flight hours are available for inspection at ________

(2) That the aircraft is not registered under the laws of any foreign country; and

(3) That legal evidence of ownership is attached or has been filed with the Federal Aviation Administration.

NOTE: If executed for co-ownership all applicants must sign. Use reverse side if necessary.

TYPE OR PRINT NAME BELOW SIGNATURE

EACH PART OF THIS APPLICATION MUST BE SIGNED IN INK.	SIGNATURE	TITLE	DATE
	Ima Loft	Owner	3/23/06
	SIGNATURE	TITLE	DATE
	SIGNATURE	TITLE	DATE

NOTE Pending receipt of the Certificate of Aircraft Registration, the aircraft may be operated for a period not in excess of 90 days, during which time the PINK copy of this application must be carried in the aircraft.

AC Form 8050-1 (3/90) (0052-00-628-9006) Supersedes Previous Edition

Fig. 9-1. *Registration application.*

white copy of the Bill of Sale, the white and green copies of the registration application, and a check for $5 (made out to "Treasurer of the United States") to:

Federal Aviation Administration

Aircraft Registration Branch, AFS-750

P O Box 25504

Oklahoma City, Oklahoma 73125.

State organizations

Most states require registration of private aircraft and levy a sales or use tax on aircraft purchases. State Registration fees vary. In my home state, they start at $50 for simple two-seaters and go on up.

If you buy a new aircraft, you'll probably have to pay sales tax (Fig. 9-2). If you buy it outside of your home state, the state "revenooers" may hit you for it when you register the aircraft with the state aviation department. In some cases, the FAA passes the information from AC Form 8050-1 to them, and they'll notify you of the need to pay a tax. If you buy the plane out of state, you should be able to pay just your home state's taxes. You should not have to pay the both states' entire tax. In any case, get itemized receipts. Sometimes your own tax liability will be decreased by the amount you paid in another state.

In most states, you don't have to pay "sales tax" if it's a private party transaction. However, you can be hit with a "use tax," which (surprise, surprise) is about the same percentage as the sales tax you don't have to pay. Depending upon location, you'll have to pay 3 to 8 percent of the selling price of the aircraft, as shown on the Bill of Sale.

Now here comes the interesting part. Sometimes, when you buy an airplane, the seller "neglects" to fill out the selling price section on the top of the Bill of Sale.

Can't turn in sloppy paperwork, can you? Better put (heh, heh) something down.

Fig. 9-2. *Most states will collect sales tax on the purchase of new airplanes like this Aviat Husky. (Courtesy Aviat Inc.)*

Be forewarned that the revenooers usually have aircraft price guidelines. A friend spent a nervous 15 minutes when the tax man challenged the amount shown on a Bill of Sale. Fortunately, the aircraft in question was a homebuilt and wasn't listed in the tax man's book. Fines and prison terms tend to interfere with your flying.

That's not to say you can't look for legal ways of reducing the bite. Some states don't levy a tax if the plane doesn't enter the state for at least 90 days after purchase. It pays to check around. Your state aviation department might be a good place to ask for specific tax provisions.

The FCC

Back in the 1990s, the FCC eliminated the requirement for licenses for radios installed in aircraft. However, this only applies to *domestic* operations. If you plan to fly to Canada or Mexico, you'll need an FCC station license for the airplane (and a restricted radiotelephone permit for yourself).

FCC Form 605 must be submitted to the FCC. It can be accessed by using " FCC aircraft radios" as an Internet search term. The hard copy of the form includes the complete instructions for filling it out and sending it to the FCC. It can also be submitted online. The FCC varies the rates a bit, but currently, an aircraft radio station license costs $105. It's good for 10 years.

While the FAA forms are surprisingly understandable (for government paperwork) the FCC application is a bit more obtuse. Take the time to fill it out carefully—some years back, FCC announced it would automatically reject all erroneous applications—and keep the money! A hundred bucks for a typo seems rather steep. A number of organizations convinced the FCC that the policy was not a good move. But hell hath no fury like a bureaucrat denied revenue, so don't tempt fate. Double-check your application.

Other paperwork

You can apply for a "vanity" registration number if you wish. Your options aren't quite as broad as those for cars; it must, of course, begin with "N," and can't be more than five additional characters long. It must consist of at least one number and end in no more than two letters.

The drawback is that the old registration number is already permanently applied. For this reason, custom "N" numbers usually are the province of the homebuilders. Still, if you are restoring the aircraft (or just repainting) you can get a custom number. AC Form 8050-64 is used.

INSURANCE

From most aircraft owners' viewpoints, the insurance world is a vast jungle studded with pitfalls and sharp-toothed carnivores. There's some truth in that. Insurance is a rather esoteric field; one of which most of us prefer to avoid (Fig. 9-3). Still, you're

Fig. 9-3. *Most pilots dislike having to pay for aircraft insurance, but it does come in handy on occasion.*

an airplane owner now. Grab a machete and be prepared to hack through some red tape.

We talked a bit about insurance in Chap. 2. Back then, I recommended you call some agents and get some price estimates. Now that you have a real piece of hardware, call them back and get definite quotes.

Before you do, let's examine the types of coverage, plus the loopholes, and whatnot that lurk within insurance policies. One thing to keep in mind: Insurance policies vary from company to company. Subtle differences in wording can make a world of difference. Quiz the agent for anything you don't understand.

Liability coverage

Typical liability insurance coverage is $100,000 per passenger in the aircraft, $100,000 per person on the ground, and $500,000 for property damage. The insurer also places a cap on how much the company will pay if you are found liable for an accident. Less coverage is certainly available, such as a $100,000 property damage limit. But the decrease in premiums isn't that much, and if you lose control on landing and roll into just about any new aircraft, the policy won't cover the total cost.

One option you'll see is the Combined Single Limit (CSL) policy, usually called a "smooth" unit. This gives you, say, $500,000 in coverage but doesn't establish sublimits for bodily injury or property damage. It may yield a lower premium, but can cause some problems if litigation follows an accident. The liability premium will vary with how many seats the airplane has. Coverage of my single-seat home-built costs $500 a year; the owner of the seven-seat Cessna Station Air in Fig. 9-4 will pay far more. If you have a six or more-passenger airplane, you can sometimes reduce your premium by removing one set of seats. Assuming that it's allowed by the aircraft's Type Certificate, of course.

Some owners operate "bare," with no liability coverage. It's a tough course to take in this litigious age. Considering all the other expenses of ownership, the cost of liability insurance isn't that significant.

Fig. 9-4. *The insurance premiums for larger aircraft can sometimes be reduced by removing seats. Cessna Aircraft Company.*

Hull

"Hull" coverage is the equivalent of automotive collision or comprehensive coverage. Some aircraft owners don't carry it. They prefer to "self-insure;" if they break the airplane, they'll fix it as their funds permit. Deciding to self-insure is like betting with the insurance company. Fly twenty years without crashing and you win! It's not for the faint of heart or those with expensive airplanes.

There are several different varieties of hull insurance, depending upon what level of protection the owner desires. These levels can allow you to get the best coverage for the available money. From the least expensive on up:

1. "Not in motion" policies cover the aircraft for physical damages, which occur when the plane isn't moving and/or the engine isn't running. This is useful for protection against weather damage or vandalism. If hail smashes the windshield or dents the wings, the insurance pays.

2. "Not in flight" policies cover the aircraft for physical damage up to the point where it taxis onto the runway for takeoff. The policy goes back into effect when the plane safely reaches the taxiway after landing. If you taxi too close to a signpost and ding a wingtip, the policy should apply.

3. "In-flight" should cover your aircraft against risks on the ground and in the air. This coverage is sometimes called "all risk," though that isn't quite accurate. All policies contain exclusions; cases where the insurance isn't valid.

The specific versions of these types of coverage vary between insurance carriers. It is vitally important to understand the "definitions" and "exclusions" listed in your own policy. The coverage you select should depend on your level of comfort. Back when I owned a $6000 Cessna, I carried "not in motion" coverage. I was willing to bet that my piloting skills were sufficient to prevent the need for an "in motion" policy.

This issue of whether or not to carry "all risk" coverage is moot if the airplane acts as loan collateral. The lending institution will require full hull insurance.

As a ballpark figure, full insurance for a Cessna 172 will run about a thousand dollars a year, depending upon pilot qualifications. Just about all of the aviation insurers offer online quote services, so it's easy enough to find out the numbers for your own particular situation.

The perils of underinsurance

There's one way to economize on hull insurance coverage. But it can really bite. Let's assume you bought a used Aviat Husky for $50,000. If you insured it for, say, $25,000, the hull premiums would be less. Moderate damage—say a prop strike or a minor ground-loop—would be covered just as well as if you carried coverage for the full $50,000 value of the plane.

But what if something really bad happens? Say a tornado hits, and crumples one wing pretty severely. It doesn't take much to run up a high repair bill. The insurance company examines the plane and determines the mangled Husky has suffered $20,000 worth of damage. That $20,000 is 80 percent of the insured value. The company will probably declare that the aircraft is a total loss. They pay you the $25,000 the policy is worth, and that's the end of that.

Twenty-five thousand dollars isn't enough to buy a replacement Husky. So you're out of an airplane, even though the amount of damage was less than half the actual value. Some salvage dealer will buy the wreck at an auction for $15,000 or so, rebuild it for another $20,000 (they've got the parts), and sell it for $35,000. Making a $15,000 profit that essentially came right out of *your* pocket.

Keep this in mind if you upgrade your airplane with new paint or an overhauled engine. These improvements increase the value of your machine and should be reflected in your hull coverage. Don't forget to have the insured amount track the actual replacement cost of the machine.

Overinsurance has its faults as well. Let's assume, for some reason, you insured that Aviat for $100,000. The premiums would be higher, but at least if the plane is totaled you'll definitely be able to buy another one right away.

Then a windstorm hits the airport and blows the Husky across the field. The adjuster looks at the results, and estimates $40,000 damage—almost what you paid for it. But the amount of damages is just 40 percent, to the insurance company. The plane isn't totaled. They could hand that steel-and-fabric tumbleweed to the lowest bidder. You'll get an airplane back months later. It probably will never be the same. And you'll never be able to sell this "major damage history" airplane for anything like you paid for it.

It won't necessarily happen that way. Some companies recognize when planes were overinsured and will negotiate a total loss figure. Still, it'll be less problems all around if you insure the plane for what its worth. Don't forget to adjust the coverage as your plane's value changes.

Policy language

Now we come to the nitty-gritty. Here are some phrases or concepts that might be included as part of your insurance policy. Don't take these as universal, though, the same terms might be used differently on your policy.

Geographic Exclusions. The policy may only be in effect when the aircraft is within a certain area. For instance, the coverage may apply only in "the 48 contiguous United States." If you wanted to fly to Canada, Mexico, or Alaska, you'd have to get an additional policy (rider or endorsement) to ensure coverage.

Named Pilot(s). Most policies are in effect only when specific pilots are flying. The remainder list the pilot qualifications necessary for the policy to be in force. If the pilot isn't named on the policy or doesn't meet the training/flight experience requirements, the policy won't be in effect. Most policies will cover pilots with commercial licenses test-flying your plane in conjunction with maintenance.

Pilot qualifications. The pilot must meet all the FAA requirements relative to flying the insured aircraft. If your medical or Biennial Flight Review (BFR) isn't current, then your aircraft policy may be void. If the aircraft is "high performance" and you don't have a high performance endorsement, the insurer might refuse a claim.

The aircraft must be airworthy. If you groundloop and the annual is overdue, the insurance company may deny your coverage. If a "not in motion" policy is in force, this exclusion shouldn't apply. Ask your agent. Better-quality policies don't include the last two provisions when the discrepancy cannot be shown to causal; that is, contributing to the accident. In other words, if your third class has expired, the insurance should be in effect, unless some medical condition caused the accident or made the outcome more severe.

Compliance with Federal Aviation Regulations. This provision is loaded with dynamite. Most accidents are accompanied by a FAR regulation. 14CFR 91.13, "careless or reckless operation" can probably be applied to just about any time aluminum is bent. The provision is tough to avoid, but at least the insurance companies don't seem to invoke it except in the more extreme cases.

Legal or Illegal Seizure. If you use your Cessna 180 to haul happy dust into the country, the insurance company doesn't compensate you if the Feds seize your airplane.

Betterment. To an insurer, "betterment" is when the repairs after an accident leave the aircraft in better shape than before. That's a no-no. Say your brakes lock on landing, blowing two tires and causing some wheel and brake damage. Your insurer will look carefully at those tires. If they'd been worn halfway prior to the accident, the insurance company may only pay for one half the value of the new tires. If your constant-speed prop hit something while the plane screeched to a halt, the same applies. Some models have to be torn down and overhauled every five years; if your prop's time was up, the insurer may refuse to pay for it.

Engines are a bit different. Good hull coverage pays for post-accident teardowns and inspections in "sudden stoppage" cases, generally regardless of engine time. And the insurance will pay for damage caused by the accident. But if the inspector finds a spalled camshaft, replacement comes out of your pocket, not the insurer's.

Here's another example: You're flying along and the engine quits. You do a semi successful forced landing. If the engine quit because the crankshaft broke, the insurance *will not cover* its repairs. That would be betterment. If the failure was due to a clogged gas vent and the engine was damaged on impact, the insurance *does* cover the engine repairs.

Breech of warranty. If you took out an aircraft loan, the lender required hull coverage to protect the loan's collateral. If you look on your insurance policy, you may see the term "breech of warranty." A breech of warranty clause requires the insurer to pay the coverage amount to the lender even if the policy is otherwise voided. So if your BFR is overdue and you overshoot the landing, the insurance company still pays off the loan.

Home free? Not hardly. The insurance company will sue you for the amount they had to pay your lender, a process called *subrogation*. Again, check over the policy thoroughly to be sure the provisions are acceptable.

The criticality of compliance

It is important to exactly conform to the insurance company's requirements. Insurance companies make money by collecting premium payments, not paying off claims. If you buy a retractable gear policy and the insurer requires "Owner must log fifteen hours of dual before hull coverage commences for solo flight," make *darn* sure that at least the required amount of time is entered in your logbook prior to flying solo. If you take your fifteenth hour of dual and forgot to bring your logbook for the instructor to enter the lesson, don't go for a celebratory solo jaunt around the pattern. Remember, the requirement is for logged time. Otherwise, you'll just be giving the insurer grounds for denying any claim.

The same situation holds true when applying for coverage. The insurer may not scrutinize your application when your coverage starts, but they may examine it closely if you file a claim. If you claimed a thousand flight hours, but your logs only show 500, that's grounds for denial.

If you're carrying "not in flight" or "not in motion" hull coverage, understand exactly the insurer's definition of these terms. "Not in motion" coverage should have some nice legalese along the lines of "Motion imparted by engine thrust or as a result of flight."

My friend Jim once had a passerby help him push his plane into the hangar. The over-muscled stranger shoved it right into one of the roof supports. If his hull coverage had only said "not in motion?" After all, the plane was in motion at the time of the accident.

If you don't understand any term on your policy, get the agent to explain it to you. Ignorance is no excuse for noncompliance.

Starting the coverage

If you work through a local agent, insurance should start upon your delivery of a check for the premium. Most companies have payment plans, so the entire amount may not be immediately due. Some companies have monthly plans, some quarterly and others semiannually. Again, it's something to ask about before the actual aircraft purchase.

To minimize the premiums, become a better pilot. Get advanced tickets, especially instrument ratings. Participate in the FAA's Wings program. If you're carrying hull coverage, put the plane in an enclosed hangar. Remember, you'll be paying quite a bit for your insurance coverage. Make sure you don't negate it by violating some term of the policy.

TYING DOWN

The plane's yours, but you can't just let it sit around loose. You'll want to tie it down before you go home that first day. Maybe you've got a hangar, and figure you can skip this part. But if the hangar is the open type, the wind will still blow through. In these cases, most airports require that planes in these hangars are tied down as well.

Most of us are going to have to secure the airplane. There are two elements to this: The anchors and the tiedowns.

Anchors

If the tiedown spot is paved, the anchors will consist of an imbedded steel loop, hook, or eyebolt. There's usually not much you can do to improve these anchors. Take a look, though; make sure they're solid.

Grass parking areas have a variety of anchor systems. Some have a long cable strung across the tiedown area, tacked down at intervals. Or, each spot might have an individual anchor. Cast a jaundiced eye at the anchors on turf. Talk to the airport manager; the tiedowns are usually the complete responsibility of the owner. Those at your spot might have been left by the previous occupant. You then have your choice of using it as-is or installing your own. A pull test might be a start, but you'd like each anchor able to withstand a thousand-pound pull. Tough to verify by just tugging on it.

Otherwise, you can install your own. You need something that goes in deep and grabs onto a wide patch of dirt. Aircraft supply catalogs sell tiedown kits, but you might do better at a local farm-supply store. You could even dig a deep post-hole, insert a hooked steel rod, and fill it with cement. You'll need anchors set apart at the approximate distance between your wing's tiedowns, plus one a few feet back of your tail. Some people use concrete blocks to tie down the tail, but I'm pretty skeptical of those.

Tiedowns

With the anchors established, you'll need something to connect them with the aircraft's tiedown points. You'll generally see one of three systems used: rope, chains, and straps.

Rope tiedowns are probably the most common. They're inexpensive to buy, easy to attach, gentle on the plane, and easy to tighten. They can have a problem with long-term exposure to the environment. I once tried to tie my plane down at a transient spot at a remote field. I grabbed onto one of the ropes, leaned backward, and it broke. So did the other two. Ultraviolet light had done them in.

You can choose between three general materials: manila, nylon, and polypropylene ("poly"). Manila will be the cheapest. It's rather poor at strength and abrasion resistance, and moisture causes deterioration. It'll stretch and shrink with the weather. Plus, you'll have to learn how to do a little bit of rope splicing to keep the ends from unraveling. However, it withstands sunlight a lot better than synthetic ropes. Nylon and poly are the most common. Poly gets a lot of attention in the boat world since it floats, but that's neither here nor there for aircraft. Poly is a bit more susceptible to abrasion, damage, and deterioration from sunlight than nylon.

You'll want about half-inch-thick rope. For nylon, that corresponds to about 4000-pounds capacity. Its "working rating" is usually 10–20 percent of its capacity. A half-inch nylon rope is good for about 600 pounds. The ends of nylon and poly rope will tend to unravel. After they're cut, get a match or other flame source. Move well away from any aircraft and light the end of the rope. Let it hang down and burn for 10 seconds or so, then blow it out. Let it cool. The end of the rope will have melted together. A car's cigarette lighter can be used instead of a flame, but it might take several applications. The lighter stinks a bit afterward, too.

You won't have to do the same trick with a chain, though you'll need a hacksaw to cut it to length instead of a penknife. You'll also need D-links or something to attach the bottom of the chain to the anchor, and a clip for securing the upper end. Chains are, at first thought, a natural for aircraft tiedowns. They're strong and don't deteriorate. However, they're tough to tighten. If the links are each an inch long, you can end up with that much slack in each chain. If a wind rocks the plane, the chain jerks it to a stop. That jerk can induce serious loads; in extreme case, you could damage the tiedown point. Rope can be adjusted to exactly the right length. And it has a little "give," which can reduce the strain in gusty winds. Chain can also scratch the aircraft's finish. Always secure the free end so that it won't swing in the wind.

Finally, there are web or strap tiedown systems. By themselves, they are similar to nylon or poly rope. Some companies sell systems with tensioners that make it easy to take the slack out. I don't really have any experience with these, but they probably work OK. It's just more expensive than rope.

How to get tiedowns tight

I can't tie a knot worth a darn.

There are some ways around this. I use a little homemade tightener; similar units are available commercially. But there is a way to ensure snug ropes without any additional hardware. All it needs is the tail anchor to be a bit behind the aircraft.

Park the airplane with the wing tiedown points right over the anchors. Make sure the parking brake is off. Tie the wings down as tight as you can. Go to the tail and put the rope through the eyelet, or around whatever other tiedown point is provided.

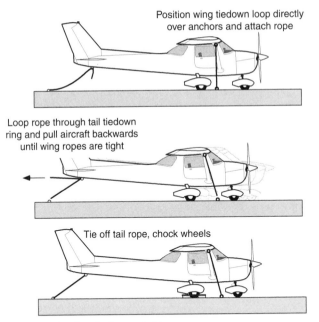

Position wing tiedown loop directly
over anchors and attach rope

Loop rope through tail tiedown
ring and pull aircraft backwards
until wing ropes are tight

Tie off tail rope, chock wheels

Fig. 9-5. *Getting tiedowns tight.*

Then pull on the rope to tug the aircraft backward. The wing ropes will tighten up. When the plane stops moving, tie off the tail rope. This process is summarized in Fig. 9-5.

AIRPORT ETIQUETTE

Until now, you've had a pretty easy flying life. The FBO probably had the plane ready on the ramp when you came out to fly. They probably handled the fueling and preparation. All you had to do is preflight, hop in, start, fly, and park it in the spot you left from. As such, you probably haven't had much of a chance to inconvenience or anger someone else. And even if you did, you were flying an anonymous rental aircraft, right?

That's changed now. You're going to be living next to a bunch of other aircraft owners. They're a swell bunch of people, mostly. They're willing to bend over backward to help a fellow owner. You'll be able to pick up a lot of advice on maintenance items, insurance, mechanics to avoid, and a lot of other subjects.

These owners will only ask one thing: That you don't do anything that might damage their own aircraft or interfere with them unnecessarily. To prevent you from unnecessary faux pax, let's discuss some airport etiquette.

Watch your &@%$^# prop blast!

I talked to a lot of pilots about what really got them mad. Number one on the list, by far, is owners who don't minimize the propeller wash hitting other aircraft. Even at idle, propellers fling a lot of air backward. Prop blast flings gravel on paint jobs. It shakes planes off jacks. It blows cowlings across ramps. It knocks over ladders. It catches aircraft doors and slams them against the stops. It seriously ticks people off. Believe me.

There are some incredibly stupid people out there. Years ago, there was a guy with a Cessna 182 just two hangars down from me. When he went to fly, he rolled the plane out of the hangar and turned it 45 degrees to the right.

The tail pointed right at the airplane in the open hangar next to him—an award-winning fiberglass homebuilt. And time after time, I'd seen the 182's owner start his engine. And sit there idling until his engine warms up. All the while, gales of wind scoured the homebuilt. Then the 182 driver would gun the throttle to start taxiing toward the runup pad.

There is absolutely no reason for such behavior. The owner could have easily pulled the plane 15 feet farther and point the tail along the taxiway. It's stupidity in this case, "hot-dogging" in others. One advantage of taildraggers is that the tailwheel unlocks and the airplane can pivot in a tight circle. With a liberal application of power, the plane fairly pirouettes. I've seen pilots do this to turn around in narrow taxiways. Sure enough, they blast about six planes in the process. And seem to think that the glances shot their way are admiring ones!

You should have the idea by now. Here are some specific rules to live by:

1. Never, *ever* start your engine with the tail pointed at someone else's airplane. Roll the plane out of its parking spot and turn the tail to point down the lane between aircraft. If it's impossible to avoid (like you're parked in a "U" of airplanes at a fly-in), roll the plane forward by hand as far as possible, first. Ask bystanders for help—most owners will be glad to assist.

2. Use no more power than is necessary. Even with the tail pointed down the taxiway, propwash spreads.

3. Use momentum when you can't avoid pointing the tail at someone. For instance, if you have to turn to the side to point the tail at your hangar, let speed build a bit and pull the mixture early. Make the final turn as the engine winds down. It may look like hot-dogging, but surrounding owners will bless you for it.

4. Finally, not every pilot likes aircraft noise. If you need to run the engine for a while to warm it up for an oil change, taxi to an isolated part of the field. Don't just roll it onto the taxiway in front of the hangar and run it up.

Don't block the hangar/taxiway!

Many airports have a rather narrow lane between tiedowns or hangars (Fig. 9-6). Yet there are always owners who narrow the space farther by the way they park their cars, or by rolling their plane partially onto the taxiway, then drive away for lunch. And even if you tuck your car off to the side, make sure you don't block anyone's hangar door from opening all the way. The person in the hangar next to me used to leave his car parked along the wall between our two doors. He'd position it so his hangar wouldn't be blocked, yet the front of his car would encroach the door of mine.

Once, another man came around the corner to find me, and a couple other guys lifting a wing over the hood of his pickup truck so the plane owner could put his

Fig. 9-6. *Most pilot are nervous when taxiing in tight quarters; don't make it worse by parking your car in the wrong spot.*

machine away. The truck owner was apologetic, but that didn't make him any more popular. And I broke my watch sliding it across the hood of his pickup.

When we drive cars, we have a pretty good feel where the fenders are. But people taxiing airplanes don't like to take risks. When you park your car, it may look like any plane has plenty of room. But things look tighter from the cockpit. Remember, the taxiway is to *taxi on*—not a place to preflight your plane, wait for passengers to arrive, or to clean the bugs off before putting your plane away. A plane sitting idle in the taxiway prevents others from passing (Fig. 9-7). If you must block the way, keep your eyes open to be ready to put your plane back if someone needs to pass.

To summarize:

1. Don't park your car where there's any possibility of impeding aircraft traffic.

2. Don't park your car directly across from another parked car or other obstruction.

3. If you're going to be gone for a while, park in your airplane's spot (assuming the airport allows this practice).

Fig. 9-7. *Don't roll your plane onto the taxiway until you're ready to start the engine.*

4. Never immobilize your plane in a position where it could interfere with traffic.

5. Don't push the plane onto the taxiway until ready to start the engine. Do your preflight at the tiedown or in the hangar.

Ask before touching

In the wild west, you didn't climb down off your horse on someone's ranch until you were invited. Similarly, don't enter someone's hangar without asking first. It doesn't take anything elaborate. "Hi! Nice plane. Mind if I take a look?" Most owners won't mind. But they'll be a little less hostile if they feel you're careful around their aircraft. Sure, if it's a big open hangar with everyone's airplanes right next to each other, you don't really have to ask just to walk up to the airplane. But for courtesy's sake, you should still ask for permission before examining closely.

It might pay off. Our club's little open-cockpit homebuilt used to sit in an open hangar, and a lot of people wandered by to look at it. I really had no problem. A number of times, I looked up from my work to see someone leaning over the wing to look into the cockpit. Again, no problem. But sometimes someone asked "Mind if I take a look at it?" first.

Those that do, get invited to sit in the cockpit. Because I figured that since they were cautious enough to ask, they probably could be trusted clambering over the cockpit sides. On a related note, though, one should *never* touch an airplane without permission. Don't wiggle the controls surfaces (the owner might be working on them), don't "plonk" the fabric (it causes "ringworm ridges"). Don't stick your nose where a lot of little goodies are lying loose; the owner might think you're trying to steal something.

Above all, don't help move an airplane without being specifically invited. Owners prefer to keep full control, especially in tight situations. Plus, you may pick the wrong place to push.

Other peeves

Keep kids and pets on leashes (Fig. 9-8). A dog that runs out to chase an airplane could get diced by the prop. And the antennas and pitot tubes jutting out from every airplane are often well-placed to get bent or broken by a running five-year-old. One man found a child riding "piggy-back" on the top of his Champ's fuselage, with the proud parent snapping a photo.

- Turn your strobes off when around hangars or tiedown areas at night. They can blind someone trying to secure or preflight their aircraft.

- When you drain fuel during preflight, don't dump it on asphalt. Asphalt is petroleum-based; gasoline eats it away. Concrete is safer, or pour the gas back into the plane if it was clean.

- Yell "clear!" so people in nearby hangars can hear you, and wait a moment to let them secure any loose components.

- Never leave a stepladder upright—it could get blown over.

Fig. 9-8. *When taking your pet human for a romp on the airport, bring pretzels and a leash.*

Uncontrolled airport guide

If you're like most pilots, you've learned to fly from a field with a tower. You undoubtedly make some flights to nearby uncontrolled fields, but most of your operations have been guided by ATC. It's quite possible, though, you may end up based from an uncontrolled field. Tiedowns and hangars are generally cheaper, and they're often convenient to the suburbs where most of us live.

There are a few points to remember around uncontrolled fields:

1. Runway space is gold, especially on sunny summer weekends. After landing leave the runway at the earliest safe turnoff. Add power to reach the next one if necessary.

2. Don't block access from the taxiway to the runway during your runup. Some aircraft have very short pretakeoff checks. Most fields have a runup area; use it.

3. Many uncontrolled airports have noise problems with the neighbors. Don't make them worse. Follow noise abatement procedures.

4. Don't abandon your plane at the gas pumps. After filling up, get clear so that other planes can fuel.

5. A small percent of aircraft in the United States don't have radios. Few of these are based from tower airports, so the proportion is higher at uncontrolled fields. Be on the lookout. Remember that they cannot hear your announcements on the UNICOM/CTAF, so avoid nonstandard procedures like 360s on final.

6. Straight-in approaches aren't prohibited by regulation, but many pilots are strongly opposed to them. You'll make no friends if you make straight-ins a habit.

Well, we survived the first week of ownership. Now, let's take at how to handle things over the long term.

10

The First Year...
and Beyond

Your new plane is going to require some involvement on your part, beyond just wiggling the controls. Whether or not you plan to involve yourself in its maintenance, there are a number of other details, which you'll have to attend to. This chapter covers issues, which will or might arise in your first year of ownership. You may not actually face some of them within your first 12 months, but they may come up at some point. We'll cover them here for convenience.

One thing that *will* arise in your first year is the annual inspection. That's covered in the next chapter.

FEATHERING THE NEST

Once you've got your plane settled into its new nest, you'll want to make that hangar or tiedown spot as homey as possible. Let's examine some of the things that'll make your flying life easier.

Chock talk

"Chocks," you say? "Why use chocks when the plane has a parking brake? Besides, the tiedowns will keep the plane from rolling."

Why not use the parking brake? Here's a good reason: Say you're in a row of hangars with a number of other aircraft and one catches fire. The fire department decides to move your plane away from the fire zone. It has nothing to do with any deep consideration of your investment; they just want to keep your loaded fuel tanks from adding to the conflagration. Firefighters undo the tiedown ropes, simple enough. Nothing blocking the wheels. Then they pull on the plane. It doesn't roll.

Do you think they're going to try figure out why? Heck no! They'll throw a log chain around your prop or a strut and drag the plane away.

Chocks are handy things. If your parking spot slopes a bit, they'll stop the plane from rolling while it's being tied down. They'll keep it from moving when you jack one side up to change a tire. And if the plane doesn't have a starter, they're just about required during hand propping.

Chocks are another item that you can spend a little or a lot of money on. There are nice lightweight travel units available through the aircraft catalogs. If you've got a table saw, you can make some units with a nice bevel. But even a cheesy handsaw can make some nice serviceable chocks from three bucks worth of lumber (Fig. 10-1).

Control locks

Hopefully, your plane came with control locks. Letting the wind knock your ailerons and elevators around wears out the system's bearings and bushings. Closed hangars solve the problem, too, but control locks are cheaper.

You usually can secure the stick or yoke with one of the seat belts. There are drawbacks, though. At least one control is usually held hard-over. It'll add to rocking; if the wind gets really strong, it could even contribute to blowing your plane over.

At a minimum, pick up a set of the manufacturer's cabin locks. These, or similar units, are available through many suppliers. The best solution is a set of external locks on the surfaces themselves. These have the added advantage of keeping any strain out of the control system. Yoke or stick locks still allow the ailerons or elevator to strain against their hinges and cables. External locks can be expensively purchased or cheaply made from lumber or plastic (Fig. 10-2). The surfaces contacting the airplane should be padded, using old carpeting, stick-on weather stripping, and the like.

Towbars

If you've got a tricycle-gear aircraft, you should get a towbar. The manual ones are relatively inexpensive. You might think you can get by without one. Some owners maneuver their Cessnas around by pushing down on the horizontal stabilizer to lift the nosewheel off the ground.

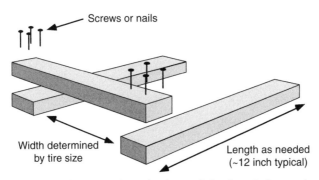

Screws or nails

Width determined by tire size

Length as needed (~12 inch typical)

Fig. 10-1. *You can make a simple set of chocks out of a couple of pieces of 2 × 4 lumber.*

Fig. 10-2. *Exterior control locks are the best for preventing damage to airplanes tied-down outside.*

Bad idea. Some older 150 and 172s have developed cracks in their stabilizer's forward spar, and ground handling has been getting some of the blame. It's not something that occurs suddenly—nothing will probably happen the first time you shove your 172 into the hangar. But long-term abuse of this sort seems to crack the spars around the bolt holes.

While a replacement spar is a seeming bargain—less than $300 at this writing—it takes quite a number of labor hours to install. The total bill for one 150 owner came to over $1,200. *Now* does that $25 towbar seem like a good idea?

If you've got a heavier plane or a steep slope at the hangar, consider a powered tow bar (Fig. 10-3). Commercial ones are expensive, but I've had friends make their own for less than $100. If you look carefully at Fig. 10-3, you'll also see another accessory that might be necessary: Small ramps to help the plane roll over the tracks for the hangar doors. While the tracks might be less than an inch high, airplane tires are relatively low diameter and it seems like they're too easily stopped by the door tracks. All you need is something to raise the wheel gradually. The photo shows metal plates; I made a small ramp out of scrap wood to accomplish the same function.

Finally, another option for pulling a heavy plane into a hangar is a small winch (Fig. 10-4). These are cheap and take little effort, but they do need to be anchored somehow to the floor or wall of the hangar.

Other aircraft accessories

How's the bird problem in your area? The feathered kind, not the metal ones. Odd now that I think of it, but in 20 years of flying I've never had a bird nest in my cowling. Probably just lucky.

Closing off the openings in the cowl helps prevent bird problems. A number of companies make custom plugs for aircraft cowlings. There are outfits who sell do-it-yourself kits, too. Or you can buy a quarter-sheet of plywood or a hunk of 3 or 4 inch

Fig. 10-3. *A powered towbar can make movement of larger air-planes much easier, expecially when the plane must be moved. Small ramps help Patrick Flynn get his Cirrus over the hangar door tracks.*

Fig. 10-4. *A winch is a low-cost method to move an airplane into a hangar, but they don't help roll it out. Plus, it must be attached to some structure in the hangar.*

foam rubber and start hacking away (Fig.10-5). One note of caution: Foam rubber does tend to disintegrate when exposed to weather. Picking little bits of foam from between your cylinders is poor entertainment, so watch your plugs' condition.

Mud-dauber wasps are good reasons for pitot and fuel vent tube covers. Again, these can be purchased ready-to-use or you can buy a little vinyl and try to figure out the family sewing machine. For aircraft with mast-type pitot systems, I've heard of one rather unusual cover: A condom. Though accidentally leaving it on might produce one heck of an embarrassing NTSB report. Be very, *very* sure you remove it before flight!

Good advice, no matter what kind of protection you use on your aircraft. The commercial cowl plugs and pitot covers usually come with long red "Remove Before Flight" banners. Add something similar to homemade units. You probably won't forget to remove windshield or instrument panel covers. These range from $5 "space blankets" thrown over the instrument panel to custom-fitted exterior covers (Fig. 10-6).

They have two major uses. First, in hotter climes, they can reduce the interior temperature. It's not just a comfort issue; your avionics and your upholstery will last longer. Second, a windshield or panel cover helps hide your expensive radios from those with a larcenous bent. Thieves won't usually break into a plane unless they know exactly what prize awaits.

Custom-fitted exterior covers are probably the all-around best. In addition to protecting your interior, they also keep the Plexiglas dirt- and dust-free between flights. But since they're on the outside, radio thieves can easily peek underneath to see what your panel's packing. External covers also take a beating from the weather and must be replaced occasionally. Then again, it's better to replace your cover than your windshield.

Speaking of beating, pay close attention to *how* the covers attach. You don't want them damaging your paint. All grommets should be isolated from the exterior; all straps should have smooth exteriors. The straps and grommets should hold the cover securely and keep it from flapping. A flapping cover can bang on the aircraft

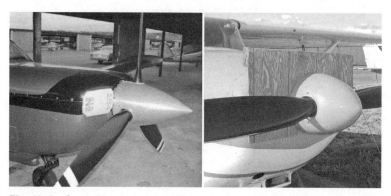

Fig. 10-5. *Cowl plugs help keep birds out of your engine. The foam kind are a bit easier on the paint.*

Fig. 10-6. *Custom-fitted cockpit covers are the best, though most expensive, choice.*

skin and rub away paint. That's one drawback to the $1.99 blue plastic tarps thrown over some planes—they can't really be cinched up tight. Still, these covers might be your only option if rainwater keeps leaking into your cabin. Just be careful of how you arrange the bungee cords (Fig. 10-7).

A second sort of cover has a shiny surface and attaches to the inside of your windows. While they don't help protect the Plexiglas, they do keep the temperature down and keep the panel hidden.

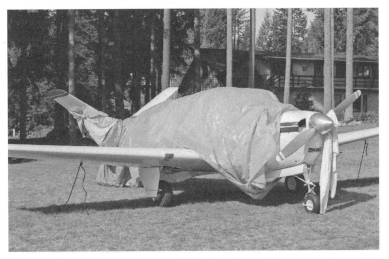

Fig. 10-7. *Impromptu covers might be better than nothing, but unless solidly secured, they can damage the aircraft's paint.*

Hangar goodies

Accessories and gadgets can make airplane ownership easier. In later chapters we'll discuss owner maintenance and required tools. Here's a few items that you may find useful now:

Engine oil. You can buy oil a lot cheaper by the case from an aviation discounter than quart-by-quart at the FBO. In any case, if the engine is low, keeping by your plane avoids a side trip.

Extra inspection panel screws. Take one of the screws that hold your inspection panels to your local aviation supplier. Buy 10 spares and put them in an old film container. If there isn't a nearby aviation hardware source, order some by mail.

A screwdriver assortment. You'll need them for the screws, right? A cheap set of screwdrivers (suitable for occasional use) can be bought for less than $10.

Plexiglas cleaner and polisher. Airspace is crowded; keep the windows clear.

Air pressure gauge. The regular automotive kind is fine, although if you've got a Cub or other plane with low-pressure tires, make sure the gauge will give readings below 10 PSI.

Tire pump. Aircraft tires don't carry much air, compared to car tires. A simple bicycle pump is sufficient for adding a bit of air to a low tire. If you're lazy like me, pick up one of those 12-volt car emergency tire pumps.

General-purpose cleaner. A spritz of household cleaner on a fresh grease spot keeps the stain from becoming permanent.

Waterless hand cleaner. Sometimes you'll get a little dirty working around the plane, and a container of "Go-Jo" or equivalent keeps you from leaving grease marks on your car's steering wheel.

A jug of water. The waterless cleaner does leave your mitts feeling a bit funny; sloshing water on them really helps.

A small glass jar or two. Something you can shoot fuel into to take a look at it, or use to test autogas for alcohol (explained later in this chapter). Old food jars are fine.

Paper towels and/or shop rags. Needed for wiping and cleaning.

Old newspapers. Use them to catch spilled oil, and other messes.

It's ideal if you can arrange a way to store this stuff out at the airport. Check on your lease and see if it's OK to place a cabinet in your hangar. Don't run off and buy a brand-new locker. Watch the garage sales and auctions. I picked up a 6-foot tall cabinet for $12 at a local surplus store (Fig. 10-8). Old school lockers work just fine, though perhaps are a bit narrow. Finally, if you have a hangar to keep it in, a bit of carpeting or a large piece of cardboard make working under the airplane more comfortable.

Don't be surprised if what appeared to be large, roomy hangar fills up pretty quickly. Since I do most of my own maintenance, my current hangar is loaded with

Fig. 10-8. *A cheap locker stored all the author's airplane goodies when he rented an open hangar. The fuel cans are empty, used only to fill the aircraft.*

workbenches and tools. I've also added creature comforts for those long weekends working on the plane—things like some comfortable chairs, an aircraft-band radio, and a small refrigerator.

In contrast, you probably won't be able to store much in an open tiedown. The airport management might allow small footlocker-like units. I've seen people use those large plastic containers designed for pickup trucks. A chain and padlock secures them to a tiedown anchor. The same hardware can secure a stepladder as well.

Portable tiedowns

You're going to be flying to all sorts of fascinating places. Some may not have tiedowns available; it'll be a good idea to put together a portable tiedown kit. You'll need three nonpermanent anchors and some rope. Don't even *think* about using straight spikes, like tent pegs. The usual anchors are large corkscrews, the longer and wider the better.

You can buy such a tiedown kit through the aviation vendors. But the same corkscrews are sold in hardware and pet stores at nonaircraft prices. I picked up three for less than $10. I screw the anchors into a piece of scrap Styrofoam and keep the whole shebang in an old GI knapsack (Fig. 10-9). When you use such anchors,

Fig. 10-9. *The author's portable tiedown kit contains three "pet anchors" and some nylon rope, stored in an army-surplus knapsack.*

screw them into the ground at a 45-degree angle. They should lean away from the aircraft's tiedown point, as shown in Fig. 10-10.

KEEPING IT CLEAN

Pride of ownership will make you want to keep your airplane clean. Getting the crud and gunk off the paint will help it last longer; especially if you wax it at the same time. Moreover, a wash-and-wax job is a perfect excuse for a detailed exterior examination

Washing

Most airports have a wash rack, which often is just a spot with a faucet and hose. Bring a trigger-type hose nozzle. Many racks have them, but they seem to disappear.

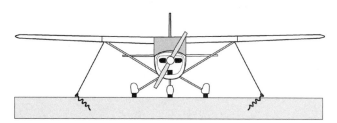

Fig. 10-10. *Portable tiedowns should be inserted at an angle as shown.*

Fig. 10-11. *Use plenty of water to avoid scratching the airplane's paint.*

Washing an airplane is similar to washing a car: Spray it to loosen the surface dirt. Scrub with a soft brush or towel with soapy water, then rinse off thoroughly (Fig. 10-11). Work in sections, rather than lathering the whole plane at once.

A sample sequence is shown in Fig. 10-12. Start with the right wing, washing it in 3-foot sections. Then divide the fuselage into four sections and do each individually: From the nose back to the leading edge, the cockpit section, the fuselage to the rudder, then the tail and rudder itself. Last comes the right-side horizontal stabilizer and elevator. On the fuselage and tail, work from high to low to chase the crud downward rather than over just-washed areas. The process repeats on the left side of the plane.

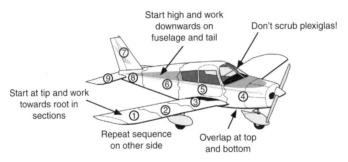

Fig. 10-12. *Typical wash sequence.*

You'll need soap or detergent in a bucket of water. Here's where a few problems may occur. Some paints, like the "wet look" polyurethanes, can be dulled by the wrong kind of cleaner. Most aviation supply houses will stock cleaners that are designed for all sorts of aircraft paint. For those whose aircraft are painted with more traditional finishes like enamel, suitable detergents are no farther away than the local auto parts store. I use a "grease release" dishwashing liquid on our club's fabric-covered airplane.

As far as applying the soapy water, just about any sort of soft brush or towel will do. Clean, of course. If you're "altitudinally challenged" (in plain English, short) or if you've a tall high-winger, a long-handled brush is mandatory. They're useful on low wingers, too; it cuts down on the leaning and stooping.

To wash a section, first blast it with the hose to knock away any loose dirt. Be careful around the pitot/static system and any engine/spinner openings, though. Get the brush sopping wet in your water/soap mixture and scrub the area clean. Re-soak the brush often to keep the area wet—a dry brush drags dirt across the paint, adding tiny scratches that dull the finish.

Speaking of scratches, leave the Plexiglas alone at this point. Hose it down, but don't scrub on it. We'll do it separately later.

How big a section should you wash at a time? Well, you don't want any dirty, soapy water to start drying on the paint. It's actually more of a problem on the fuselage, since most of the water runs away quickly. You want to rinse it clean with a blast from the hose before the dirty film has a chance to dry. It'll take a bit of technique around the top and bottom. On the top, there's always a possibility that some of the wash water will roll down the other side, which may have already been washed. Check often, and give any streaks a blast of the hose.

Plexiglas

The primary secret of Plexiglas cleaning is water, lots of it. You want to wash away any fine grit before it has a chance to scratch the plastic. When you wash the rest of the plane, blast the windows with the hose but don't touch them with the scrub brush.

When you're done with the main washing, dump out your bucket, flush it with the hose, and fill it with fresh water. Get a nice soft *clean* sponge. Soak the sponge, then gently wipe it down the Plexiglas. Don't scrub it in a circle, don't rub it back and forth. Just start at the top and slide it gently down to the bottom. Shift to the side, just enough, so there's a little overlap with the previous swath, and repeat.

When done, hit the windows with a Plexiglas cleaner. Most people use the stuff available through any aviation supplier. Stick your nose close to some windshields, though, and you'll catch a scent of citrus—some owners swear by "Lemon Pledge." Your call.

The main point is to make sure everything is clean. Set aside a sponge just for cleaning the glass. Mind your hands, too. I've never been able to wash an airplane without getting three-quarters soaked. Wet skin or clothing tends to pick up any dirt or sand it comes in contact with, so clean off before doing the windows.

Some people say the interaction between paper towels and Plexiglas causes a buildup of static electricity. This charge then tends to attract dust. Rubbing a paper towel on the windshield does seem to build a charge. Pilot outlets sell static-free wipers for five bucks or so for 150 sheets. Others use disposable diapers. I "liberate" old towels from the linen closet when my wife isn't looking.

The belly

I have a terrible confession. I generally wash the belly with the rest of the airplane, but make no special effort to scrub it clean. It haunts me at annual time. I end up flat on my back, cursing and scrubbing to get a year's accumulation of oily dirt off the belly.

There are a couple of alleviating factors in my case. First, the exhaust pipe of my Fly Baby is long enough that it doesn't stain the fuselage. Second, the crankcase breather tube (a primary source of belly oil) is positioned to reduce the problem. With a production airplane, you're pretty much stuck with the way Mr. Cessna or Mr. Piper sold it.

You can buy aircraft cleaners specifically designed for cleaning exhaust stains and such from the belly. Again, find one compatible with the aircraft's paint. Be sure to follow the dilution instructions. Some of these products must be diluted 20 to 1 before use! Sure, it'll clean better at the higher concentration. But stripping it down to bare aluminum is probably not your first choice.

The instructions generally call for these products to be sprayed on and allowed to sit for a while. Garden shops sell pump-type sprayers, or you can be satisfied with the rather anemic output of the grocery-store spray bottle. Whichever type you use, *read the label first*. Some caution against using them on aluminum.

Other

Landing gear parts, such as oleos, tend to have a little seepage of hydraulic fluid; this makes them dirt magnets. The grit chews on O-rings and seals, making the problem worse. Clean off the greasy parts with your cleaner. If it isn't quite strong enough, try petroleum solvent, mineral spirits, or a grease cutter called Varsol. Gasoline works too, but it's not a safe practice. If you must, wad up a paper towel, hold it under one of your fuel drains, and put a quick shot of gas on the towel.

The interior can be cleaned just like a car. Aircraft carpeting is generally designed to be easily removable (it usually has to come out at annual time), so it's easy to vacuum up the dirt. Be careful when using cleaners in the cockpit; the windows are Plexiglas on the *inside*, too.

Waxing

Waxing the paint makes it look better and last longer. I've used automotive wax with acceptable results, though there are some deluxe silicone aircraft polishes that aren't really that much more expensive. The car-type waxes that I've used tend to get caught in seams when wiped on; then I have to dig it out.

If you've got a fabric-covered bird, be careful before you try traditional wax. The residue can get lodged in the fabric weave and be the very devil to get rid of. It isn't a safety issue (the weave just happens to show through the paint) but looks like hell. Try it on a small, out-of-the-way area first.

Engine cleaning

Cleaning the engine isn't as simple as it once was. It used to be that you could take off the cowling, wrap plastic bags around the electrical stuff (magnetos, regulator, and alternator) and vacuum pump, and blast away with engine degunker.

Environmental regulations are starting to limit the practice. My home field won't let us clean the engines at the wash rack anymore, and you really would like to flush the engine down with water after the degreaser does its thing. If your airport isn't as restrictive, you can use the engine cleaners and solvents sold at auto-parts stores. A pump-type sprayer puts it on. Let it sit for a bit, then flush it with clean water.

Due to my airport's limitations, I've taken to using spray cleaner and paper towels. It's slow and tough to get into the tight corners, but it works. The process also gives me a slow, detailed look at the condition of the engine. Cleaning the engine is also required at annual time. If you fly only 50 or so hours every year, your engine probably won't get dirty enough between annuals to require more than just a bit of wiping down.

SHARING HANGARS

If you own a simple aircraft like a Cherokee or 172, hangar rent can represent two-thirds of your fixed costs. That is, assuming you can find a hangar. Sure, an outside tiedown is all right for an all-metal bird. But what about rag wing airplanes or that water leak you just haven't be able to eliminate?

Real birds share nests—why not aluminum ones? Squeeze two planes into a single hangar (Fig. 10-13), and you can slice your fixed costs by a major percentage. Your pride and joy can stay warm and dry for not much more than the cost of an open tiedown.

It's not as unwieldy as you might think. General aviation hangars are oversized in order to fit a wide variety of planes. A pair of two-seaters, or a 172-sized plane with a homebuilt, probably can be made to fit. It may amount to a three-dimensional jigsaw puzzle, but careful planning and trial-and-error will yield a solution in a

Fig. 10-13. *The owners of this Cessna and Piper have both reduced their hangar cost by 50%.*

surprising number of cases. It depends on your willingness to put up with a little hassle in order to save money.

The legalities

Before you get your hopes up, check which local airports allows sharing. Some leases and city ordinances allow only a single aircraft per hangar. Some allow sharing, but require the second airplane to pay the equivalent of a monthly tiedown bill.

Finding a prospect

If you're OK legally, a prospective partner may be found via airport bulletin boards. If you already have a hangar, check with the names near the bottom of the airport's waiting list.

Try to select a partner with an airplane that maximizes your chance for success. Overlapping configurations and good ground maneuverability are key ingredients. Opposite wing arrangements (i.e., one high wing and one low wing) are best. Conventional gear aircraft are easier to maneuver than trigears. Check with your local EAA chapter, too. Homebuilts are typically smaller than production aircraft (Fig. 10-14), and their owners are more used to handling aircraft in tight quarters like basements and garages.

The basic arrangement of aircraft in the hangar depends upon many factors. Who flies the most? That plane should be closest to the door. Which is the hardest to handle? The plane in the back of the hangar generally requires the least maneuvering. An owner of a pristine show plane may prefer to keep his or her plane in back to minimize handling.

Making them fit

Determine the ground rules with your new partner, then gather some buddies. Extra hands prevent scratches and dings during the trial-and-error period. Keep in mind, though, that they won't be around for everyday operations. All aircraft movements must be executable by a single person.

Fig. 10-14. *Homebuilts are good candidates for hangar partners, because they're usually smaller than factory-built airplanes.*

Before beginning, move both props to the horizontal position. Have two sets of chocks handy. Also, every helper should carry a short piece of 2-by-4 as an emergency chock. Your basic plan is to get the first airplane (the "back" plane) all the way into a rear corner of the hangar, then bring the other plane (the "front" plane) along the other wall. Roll the first airplane into the hangar with the wingtip close to the left wall until the rudder is a few inches from the back. Note that the tip of the horizontal stabilizer is still 5 feet or so from the side wall. Push the tail toward that wall to swing the right wing further back. This yaw motion is easy with a taildragger, and a nosewheel can be lifted with a little down pressure on the aft fuselage (not the horizontal stabilizer). You've gained a couple inches; move the plane back some more.

Now bring the front airplane in, along the opposite wall. Watch the points of closest approach; keep the two planes from swapping paint. This is where the extra help comes in handy. They should be prepared to stop the plane instantly. If the hangar door clears the front plane's spinner, you've got it made. When they don't fit, it's usually because the left wing of the front airplane got too close to other plane's fuselage. Reposition the front plane outside the hangar. Roll it back again, yawing to the right as necessary to delay left wing contact. Various combinations of yaw may yield just enough space. I once kept a Citabria in a barn with a door narrower than the plane's wingspan. All it took was a see-sawing motion.

Some unnoticed characteristic of the front airplane may actually make it more suited to park in back. Try ideas as they occur; solutions seem to appear magically. Afterward, you'll shake your head and wonder why it took so long to see it.

Getting tricky

At some point, some sweaty helper may say, "Y'know, if we only had another inch or two. . . ." It's time to get tricky. You've moved the planes in yaw, now try pitch and roll. For instance, say you've rolled a high-wing taildragger into the hangar, and the trailing edge of the wing is a bit too low to let the other plane's rudder pass underneath. Lift the tailwheel of the taildragger a foot, and you'll gain 4 inches at the trailing edge. Place the tail on a stool, paint can, or box, and roll the other plane past.

Does a wingtip come too close to the top or bottom of the other plane? Try a ramp. Raising a wheel by an inch can drop the opposite wingtip 6 inches. Put one end of a 6-foot board atop a 4-by-4 block, and you've got a gradual ramp that'll give either wingtip at least 12 inches more clearance. It takes surprisingly little effort to pull a plane up such a shallow-angled ramp. Large-diameter or low-pressure tires roll right onto the blunt end of a 2-inch-thick board. Otherwise, bevel the end of the ramp with a circular or table saw. If you use nails to hold it together, leave a hammer in the hangar for bashing any that start to work out.

Try combinations of positions, ramps, and supports as you think of them. It took 90 minutes to figure out how to put our club's Fly Baby and a Vari-Eze homebuilt into the same hangar. If the fates smile, you'll also find a combination that works.

Day-to-day operations

Take a few moments to thank and slake the thirst of the crew. Now work out how you're going to get the planes in and out on a day-by-day basis.

A good guide is a stripe on the floor showing the desired wheel path. Use paint or tape to delineate the maximum allowable excursion for one wheel. If it crosses the line, stop the plane, because something's going to hit. For added protection, make flag stands from metal or plastic tubing. Arrange the flags so the plane will tap one before it can strike an obstruction. If a flag moves, you've got a potential clearance problem. Make sure required towbars are accessible to each owner. If a tight fit requires a nosewheel or a nonswivelable tailwheel to go sideways, set it on a trolley with casters.

Practice with help until you are used to the process. It becomes second nature with good guides and the right choreography. The Vari-Eze and our Fly Baby went together in seconds.

The down side

There are a few drawbacks. The obvious one is "hangar rash." Pad areas with tight clearances. Old couch cushions work well. You also must be willing to trust the other owner with moving your plane around. Face it, some aren't as careful as others.

Cockpit access can be awkward with both planes in place, so install control locks and cockpit covers before pushing the planes into the hangar. The restricted space also complicates routine maintenance. And if the hangar isn't enclosed, you may have to install additional tiedowns to match the stored positions of the aircraft.

But most folks can put up with a bit of hassle to save hundreds of dollars a year. While everyone's ideal is a nice, large hangar of their very own, waiting lists and budget-busting rents are the all-too-often reality. Find the proper partner, and your bird can sit snug and dry for little more than the cost of an outside tiedown.

AUTOGAS BASICS

A great way to fly more cheaply is to burn autogas in your airplane. Let's look at some of the issues involved.

Is it safe to use?

Autogas has a curious reputation in the aviation world.

Some pilots and mechanics swear by it; others caution against it. It probably varies with the aircraft. While I owned my 150, I had one engine failure, and *that* was due to contaminated aviation fuel.

I flew the 150 on autogas at least 50 percent of the time. Our club homebuilt operated on a near-exclusive diet of unleaded regular for eight years without a hitch.

There are a couple of potential problem areas for new autogas users. First, some carburetors have floats made of composite materials. These materials can sometimes absorb autogas, which waterlogs the float and causes a too-rich condition. Second, some carburetors have used neoprene-tipped needle valves; these also have shown some susceptibility to variation in the fuel. The same is true of other rubber parts in the fuel flow (such as those in the gascolator) and the varnish on old-fashioned cork-type fuel floats.

The interesting thing about these problems is that they can also be caused by exposure to 100LL aviation fuel. So if your plane has been burning 100LL, the carb and fuel system are probably OK. In any case, these problems can be corrected without major disassembly or rebuild. See your A&P.

Homebuilts often use special compounds to coat and seal the inside of their fuel tanks. Sometimes, this material can be attacked by some components of auto fuel. Check with the sealant manufacturer; monitor the fuel strainer for signs the material might be deteriorating.

Finally, some aircraft types just seem to tolerate autogas better than others. Talk with other owners and check with the type clubs. The original 80-octane avgas was a leaded fuel, some engines need the lead more than others. Since 100LL fuel has four times the lead of 80-octane avgas, filling with 100LL, every fourth fill or so is one way to ensure the engine still gets a little dose of lead. Some pilots with aircraft having multiple fuel tanks hedge their bets even further: They fill one with autogas and the other with aviation fuel. They take off and land on the avgas, and cruise on car gas. A nice compromise.

The vapor pressure is different between autogas and aviation fuel—there is apparently a higher potential for vapor lock on very hot days. This does lead to a weird but harmless side-effect. During cold weather, atomized gasoline has a tendency to condense on the inside of the intake manifold. It doesn't affect operations, but the recondensed gas tends to dribble downhill and drip out of the carburetor. It happens if the engine stops after running for just a short while. Don't sweat it unless the gas seems to keep flowing; it might then be a stuck carb float.

Getting approval

To run autogas in your airplane, you'll have to purchase an STC from either EAA or Peterson Aviation. To be eligible, both the airframe and powerplant must receive an STC.

Why both? Simple: The airframe must be approved for storing and delivering fuel, and the engine must be approved to burn it. Generally speaking, if an aircraft was originally certified to run on 80-octane aviation fuel, both EAA and Peterson have auto fuel STCs available for it. In addition, Peterson has some STCs for aircraft certified for higher octane fuel as well. These usually require some sort of modifications; it may or may not be economical in your case.

Your first step is to purchase the STC from either vendor. Contact:

EAA Flight Research Center

Auto Fuel STC

PO Box 3065

Oshkosh, WI 54903-3065

(920) 426-4843

Or:

Peterson Aviation, Inc.

984 K Road

Minden, Nebraska 68959

phone 308/832-2050

They'll be happy to take your credit card order over the phone. In addition to your card number, have the following information ready:

- Aircraft make and model
- Aircraft serial number
- Aircraft registration number ("N" number)
- Engine make and model
- Engine serial number
- Number of fuel caps

They'll send you a copy of the STC, plus a set of labels to replace the ones by your fuel tanks. Your old ones say something like, "Service this aircraft with 80-octane Aviation Gasoline." The new ones are similar, but add: "or 87 Octane automobile fuel conforming to ASTM D-4814."

Take the paperwork and the labels to your A&P. Typically, he or she will execute an FAA Form 337, apply the labels, and make the appropriate entry into your airframe and engine logbooks. As of that point, you're free to use autogas. That's it? Some of you might be wondering. Why not just use the stuff, and not buy the STC?

First, FBOs who carry autogas won't sell it to you unless the plane has the official stickers by the fuel caps (Fig. 10-15). Second, it's illegal. If an FAA inspector

Fig. 10-15. *While some airports sell auto fuel to aircraft, the planes must hold the appropriate STC.*

walks by while you're fueling and asks to see your STC, you're hosed. Third, it's not always a case of just applying the sticker. On some airplanes, the idle speed or idle mixture must be adjusted, by the STC. Fourth, not having the STC may negate any hull insurance you may carry. If you have an engine failure-caused accident and the NTSB finds auto fuel in your tanks, the insurer may be well within their rights to deny coverage if you don't have a legally-executed STC.

The right stuff

If you look at the fine print of your STC, you'll see a line requiring you to use gasoline that meets American Society of Testing Materials (ASTM) standards D-439 or D-4814. A few states don't require compliance. But even if your state is one of them, the fuel sold there probably follows the standard, anyway.

Autogas changes with the season. The refiners alter the formula slightly in the winter, for instance, to aid starting. They change it again during the summer to inhibit vapor lock. It's important to buy the *right* gas—you don't want to run "winter" gas in the summer!

How do you tell? You can't, really. But you can decrease your risk by buying your autogas from a major-brand retailer. "Honest Ed's" down the street might have it for five cents a gallon cheaper, but they might operate by buying up leftover "winter" gas in the springtime. Also, buy your gas from a popular location. A place that sells a lot of gas gets its resupply more often. The gas will be fresher.

Many states and metropolitan areas require the use of oxygenated fuels to reduce pollution. Refineries oxygenate their fuel in two ways: by adding Methyl Tertiary Butyl Ether (MTBE), Ethyl Tertiary Butyl Ether (ETBE), or ethanol, a type of alcohol. If MTBE or ETBE is used in your area, no problem. The FAA has determined that they don't cause any problems and has okayed their use in aircraft holding auto fuel STCs.

Ethanol, now, that's different. Some aircraft's fuel systems can't stand the stuff. Its usual M.O. is to cause the swelling of any rubber compound-gaskets, seals, needle tips, and the like. It has a greater tendency for vapor lock and can cause engine damage from high combustion temperatures.

Automakers understand that fuel quality can vary, and design their systems accordingly. Cars are designed to be insensitive to alcohol in the fuel. But you cannot put gas with ethanol in your fuel tanks under the terms of your STC.

But how do you tell?

Testing

Fortunately, testing for ethanol is simple. As a type of alcohol, it has a great affinity for water. You'll need a small clear container. A test tube or gradated cylinder will do; you can often buy them at hobby shops (look for the chemistry display) or photo supply stores. Whatever kind of container you use, it must come with a stopper for the top. It doesn't need to be big, but the taller it is, the easier the results will be to read.

Take your container and fill it 10 percent full of tap water. If you use a gradated cylinder, great—just remember (or mark), which line you filled to. On unmarked containers, put a line on the outside of the jar to mark the top of the water. Use a crayon, grease pencil, or transparency marker. Now, fill the container with the fuel to be tested. You'll notice that the demarcation point between fuel and water is readily visible. Cap the container, shake thoroughly, then let it stand at least 3 minutes, or until the fuel looks normal again. Check the demarcation point. If it's the same as before, great. There's no alcohol in the fuel. But if there is apparently more water than there was at the beginning, there's ethanol in your fuel. *Don't put it in your airplane!*

Alcohol combines with water a lot more readily than gasoline; in your sample container, ethanol has separated from the gas and joined with the water. This testing procedure is summarized in Fig. 10-16. You can also buy a volatility tester to determine your fuel's susceptibility to vapor lock. It's available through Peterson Aviation for about $65.

Passing autogas

By far, the biggest problem with using auto fuel in aircraft is the logistics. If a local FBO doesn't sell car gas, you'll have to haul it yourself from a local gas station. Then you've got to get it out of whatever containers you have and into your aircraft.

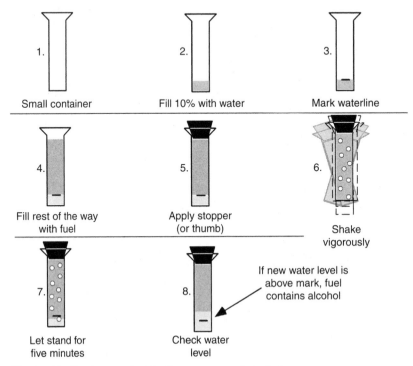

Fig. 10-16. *Testing autofuel for the presence of alcohol.*

The type of container depends upon how much gas you think you'll use. I used a 5-gallon and a 2-gallon can for fueling my 150. Since most of my flights were just for local sightseeing or touch-and-goes, 7 gallons of capacity worked well. Besides, the gas station was right across the street.

Five-gallon cans are the largest size one can conveniently handle. Transporting any more than two or three gets unwieldy. Several folks at my airport use larger containers semipermanently installed in pickup trucks. Some full-size pickups have separate "saddle" tanks. Those might be a good possibility.

Most people seem to get by with a couple of 5-gallon drums. Get metal ones or plastic ones which are UL-approved for carrying gasoline. Another thing you'll need is a grounding wire. Ten or fifteen feet of 18-gauge or bigger stranded copper wire will do. Attach a large alligator clip at both ends. Before starting fueling, clip one end to your gas can. Then attach the other end to a well-grounded point on the aircraft to equalize the static charge between the can and the plane.

Connect them in the above sequence. Don't do it the other way around. A spark may jump—it's a lot better if it jumps from the wire to the airplane than to a full tank of gas. As far as transferring fuel, getting the gas into the cans is easy enough. Getting it into the aircraft can be done in using gravity or some sort of pump.

I use gravity, but never have really liked it. I put a funnel in the tank filler, then hoist the can to pour gas into the tank (Fig. 10-17). It's not all that easy. A full 5-gallon can weighs over 30 pounds. It's unsteady, too. With my current plane,

Fig. 10-17. *The author pouring the last dregs of fuel into Fly Baby N500F. Filling the tanks by pouring directly into the tank is more difficult with a high-wing airplane that has wing tanks.*

I can do it while standing on the ground. My 150 required a stepladder. I really felt uncomfortable hauling 30 pounds of gasoline 8 feet into the air. Instead, I dumped the contents of the 2-gallon can into the tank first, then refilled the small container from the 5-gallon job.

On my current Fly Baby, I solved the problem by buying two 2.5-gallon cans. They're light enough to hoist up and pour, and the two of them fit nicely into a plastic crate so I can haul them to the gas station in my car's truck.

Depending on your aircraft and other aspects, you might be able to finesse the situation. The most popular setup I've seen lately uses a hose attachment to a standard plastic gas can (Fig. 10-18). The hose-end includes a stainless-steel ball valve, bought from the local hardware store, to turn the fuel on and off. With a protective mat on the wing, the container can be placed next to the tank. The fuel can flow either by laying the container on its side so that the hose outlet is close to the bottom, or by using a common siphon hose.

Alternately, you can rig up some sort of pump. You can find gasoline hand pumps through several sources. Build a plywood rack to hold both the pump and a gasoline can. A couple of folks at my home field have rigged up transfer systems using boat

Fig. 10-18. *An increasingly popular solution. The container can be set on the wing next to the tank cap and laid on its side, inducing fuel to flow by gravity.*

bilge pumps. They empty a 5-gallon can in just a few cycles of the handle. I had some concerns about the pump's rubber diaphragms deteriorating from contact with gasoline, but it hasn't happened after a number of years. I suspect the pumps may be designed to resist fuel or oil, since either can be present in a boat's bilge.

You can also rig up a fuel tank in the back of a pickup (Fig. 10-19). These rigs are popular, but they can cut down the utility of the truck. Do you want to drive to work every day with 200 gallons of fuel sloshing behind you? Do you want to park that huge fire hazard in your garage every night, or leave it outside where it becomes a theft target?

The ultimate solution is an electric transfer pump. These sell for about $250; you'll want the kind that run on 12 Volts DC, so you don't have to run an extension cord out to your tiedown spot. They even include genuine fuel nozzles, the kind that shut off the flow when the tank is full.

I shouldn't have to caution you about trying to rig up your own electric pump; the dangers of sparks and fuel vapors should be obvious. One bad idea some people get is to use compressed air to transfer the gas. One problem: it actually *increases* the danger. Compressing the tank raises the fuel-air mixture closer to optimal, just like inside an engine cylinder during compression.

When shopping for pumps, consider the transfer rate as well. The local hardware store sells little hand-operated pumps; these may look good but only have a 1/4-inch tube. It'd take a long time to move 30 gallons of gas. Something using 3/4- or 1-inch hose is more in the right line.

Besides transferring the fuel, you should rig up some way to filter it to prevent dirt or water from entering your plane's tanks. If you're pouring gas by hand, you'll need a funnel anyway. Aviation suppliers sell funnels specifically for fueling aircraft; these include filters to eliminate dirt and even water. With a pump arrangement, you could rig up an in-line filter. These can be found at larger hardware stores or farm stores. The filters are very much like the gascolator in your aircraft; designed to separate water and dirt.

Fig. 10-19. *A fuel tank installed in the back of a pickup truck. Having the tank decreases the utility of the truck, though, and outdoor storage can lead to vandalism or theft of fuel.*

A FINAL FUELING WARNING

Whether you use aviation or automotive fuel in your aircraft, at some point, you're probably going to be pumping it yourself. While it seems no different from putting gas in your car, there's a problem many pilots aren't aware of. Incorrect technique could cause hundreds of dollars in damage; it could even bring your airplane down.

Most fuel tanks have a short bit of filler pipe running from the tank proper to the surface of the wing or top of the fuselage. This filler pipe is not designed to support the weight of the nozzle and hose of the fueling system. As shown in Fig. 10-20, the nozzle should never be allowed to apply pressure to the pipe. The typical mistake is to place the nozzle in the pipe and allow the weight of the hose to hold the nozzle in place. The localized pressure this generates can crack the pipe, possibly allowing fuel to spill into your cockpit or wing during flight.

Don't wedge the nozzle into the tank. As Fig. 10-21 shows, run the hose over your shoulder, so that the weight cannot be applied to the pipe. Never position the nozzle, so that it's able to maintain its position in the pipe.

CASE STUDY: THE INNKEEPERS' PATHFINDER

Jay Honeck of Iowa City, Iowa was working 70 hours a week at his own business when the local airport hotel came up for sale. He and his wife, Mary (Fig. 10-22),

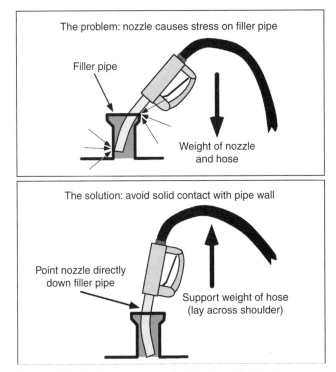

Fig. 10-20. *Don't damage your fuel tanks while filling!*

Fig. 10-21. *Drape the hose over your shoulder to keep from damaging the filler neck.*

bought the place, and they've since turned the Alexis Park Inn and Suites (www.alexisparkinn.com) into the only aviation-themed hotel in the Midwest. It not only gave them the excuse to buy aviation collectables, but they now work just 30 seconds from where "Atlas," their 1974 Piper Pathfinder, is hangared.

Atlas is their second airplane. They'd previously taken a rough old trainer, a 1975 Piper Warrior, and restored it from the ground up. But the Honeck kids were becoming teenagers, and the family needed a true four-seat aircraft. The Pathfinder, with its beefy Lycoming O-540 and 1400-pound useful load, fit the bill nicely.

Jay found Piper N56993 (Fig. 10-23) by word of mouth. "As with all great planes, it was never advertised—you just had to be 'plugged into' the 'airport network,' he says. "The best planes are *never* advertised."

"Airport network" or no, the Honecks took the usual precaution of a prebuy inspection. A good thing. The A&P found metal particles in the engine oil. Something was wearing or deteriorating, and that meant the Lycoming needed a rebuild. The Honecks bought the Pathfinder for $67,000 and put about $20,000 into the overhaul. The down payment came from the sale of their Warrior and some savings, and they took out a loan for the rest. Their monthly payment is $288.

Fig. 10-22. *Jay and Mary Honeck (also a pilot) own the aviation-themed Alexis Park Inn and Suites in Iowa City, Iowa. (Courtesy Jay Honeck.)*

Ownership experience

The Pathfinder's previous owner had added nearly every available speed modification to the plane. "We honestly get 142 knots at 23-squared," says Jay. "I think it's just about the fastest fixed-gear Cherokee flying—we will walk away from an Arrow, and give early Mooneys and Bonanzas a run for the money." The Honecks fly the

Fig. 10-23. *The Honecks named their Piper Pathfinder "Atlas" for its load-carrying ability. (Courtesy Jay Honeck.)*

airplane between 150 to 200 hours per year, with both business and pleasure trips all over the United States.

Back home, their hangar rental is $115 a month, and insurance comes to about $1,200 a year, with full coverage. "Our yearly maintenance costs have been wildly variable, depending on what goodies I add each year," says Jay Honeck. One component that he keeps a special eye on are the fiberglass wing tip tanks, a chronic source of problem in Pathfinders. "The fiberglass around the tank filler spouts degrades. I've had it done once, and it's been done several times over the life of the plane."

With its fresh rebuild, the Lycoming has been mostly trouble-free in the three years since its purchase. Jay Honeck says they made some right choices on the engine overhaul. "Because we rebuilt the engine with cylinders and pistons from Superior Airparts, we have dodged all of the O-540 ADs . . . knock on wood!" Superior Airparts makes TSO'd replacement parts for many Continental and Lycoming engines.

Jay performs the oil changes, except in the depth of the Iowa winters. Their owner-assisted annuals have run between $900 and $1,200.

Like the Warrior before it, the Honecks have spent some time and money to upgrade the Pathfinder. "We've added wingtip landing lights (a huge improvement), a new leather interior, new side panels and carpet, upgraded engine instruments, the CD player, and replaced some plastic."

An auto fuel fan

While the use of automobile gasoline in aircraft is subject to some criticism, you'll get none from Jay Honeck: "I would burn mogas in my plane even if it cost *more* than avgas," he insists.

In their first 30 months of ownership, the Honecks have burned over 5,000 gallons of auto fuel. "Atlas simply runs best on regular, 87-octane unleaded car gasoline. In fact, the only time we've had trouble with our new engine has been when we were forced to buy 100LL avgas, which causes lead-fouling of our spark plugs at the drop of a hat."

The big Lycoming is fairly thirsty, and with the Pathfinder carrying up to 84 gallons of fuel, they needed a good way to transfer fuel. A magazine article about building your own fuel trailer caught Jay's eye, but he decided to go one better. "Our new hotel needed a 'knock-around' pick-up truck for hauling furniture and stuff, anyway," recalls Jay Honeck. "So I realized that we could kill two birds with this stone, and install the tank in the back of the truck instead."

He eventually bought a "beater" Nissan pickup for less than $2,000. It was hail-dented, scarred, and painted a rather appalling purple, but it was a fairly late model and in good mechanical shape. Dubbed "The Mighty Grape," the truck received about $600 worth of work before it was deemed ready for the fuel tank.

Despite being in the midst of farm country, Honeck couldn't find a tank that would fit crossways in the back of the Nissan. Eventually, he contacted Petroleum Equipment Supply in Marion, Iowa. The company not only built a tank to fit the

Fig. 10-24. *The "Mighty Grape" has dispensed over 5,000 gallons of auto fuel to the Honecks' Piper. (Courtesy Jay Honeck.)*

Honeck's truck, but didn't charge any more than the ready-made ones in the farm supply stores.

Wiring the new tank was more complex than Honeck was ready for. After an abortive attempt on his own, he brought it to a buddy who owned an automobile repair shop. "It cost me a case of beer, a fair hangover, and a future flight, but the installation is perfect," says Jay. "I can rest easy that the thing won't start arcing when I turn it on!" A grounding-wire reel and a cheap used topper completed the fueling truck (Fig. 10-24). All told, "The Mighty Grape" cost the Honecks about $3,200.

Worth the money? The Honecks figure they saved at least a dollar a gallon on fuel. That 5000 gallons dispensed by "The Mighty Grape" earned back its acquisition cost a long time ago.

A summary of the Honeck's ownership costs is presented in Fig. 10-25.

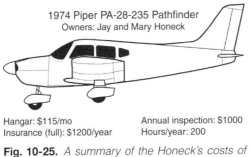

1974 Piper PA-28-235 Pathfinder
Owners: Jay and Mary Honeck

Hangar: $115/mo   Annual inspection: $1000
Insurance (full): $1200/year   Hours/year: 200

Fig. 10-25. *A summary of the Honeck's costs of ownership.*

Advice

Jay cautions first-time owners that the costs are somewhat unpredictable. "Budget 'X' dollars for ownership, but don't expect that figure to be 'hard and fast'," he says. "Some years it will be way more, others it will be less—but it's always more than you want it to be."

There's no question that the Honecks are sold on autogas. "I will never buy a plane that doesn't have the auto gas STC," says Jay. "The autogas STC is undoubtedly the best thing that has ever happened to aircraft owners."

11

Keeping it Legal

Did you realize you were taking the risk of a FAA regulation violation every time you rented an aircraft?

Take a look at 14CFR Part 91, Subpart E: Maintenance, Preventative Maintenance, and Alterations. Note how nearly every section includes either: "The owner or operator of an aircraft . . ." or "No person may operate an aircraft. . . ." In other words, if the FAA ramp-checked you, and the aircraft's annual inspection was overdue, you'd take the rap. Did you check the aircraft's maintenance logs before each flight? Did you ensure all parts replaced were legal? Did you verify that any modifications had been done in accordance with the regulations?

Probably not. Yet as the pilot, the responsibility was yours. Now that you own your own plane, the risk is greatly decreased. However, the responsibility is still there. Now, of course, the full force rests on you.

INSPECTIONS

What inspections does your aircraft require to be legal? There are four primary types:

1. 100-hour inspections
2. Instrument/equipment calibration/inspections
3. AD-required inspections
4. Annual inspection

The annual is a major enough subject to need an entire section to itself. For now, let's consider the other three.

100-hour inspections

As we discussed in Chap. 2, 100-hour inspections are required only for aircraft operated for hire (Fig. 11-1). They won't apply to you, unless you operate the plane commercially or have it on leaseback to the local FBO.

Fig. 11-1. *This Beech B36TC Bonanza only requires an annual inspection if operated privately. However, if on leaseback or otherwise used commercially, 100-hour inspections are necessary. (Courtesy Beech Aircraft Corporation.)*

A good thing, from the cost point of view. There's little difference between an annual and a 100-hour, as defined in Part 43 Appendix D. The only advantage a 100-hour has is that it can be performed by an ordinary A&P—an Inspection Authorization(IA) isn't required.

Even though it isn't required, some feel that a 100-hour *should* be done, just as a matter of safety. The decision is yours. Unless you fly over 200 hours a year, I wouldn't even consider it. Even then, it would depend on the complexity, history, and general reliability of your aircraft.

Instrument/equipment inspections

Some equipment needs separate inspections and checks at intervals. 14CFR 91.413 requires that the aircraft's transponder undergo test and inspection every 24 calendar months. If you fly IFR, 14CFR 91.411 requires an inspection and check of your altimeter and static system at the same interval.

It'd be most convenient to schedule these tests in conjunction with an annual inspection. However, your garden-variety mechanic isn't authorized to perform some of them; they are reserved for specialists at certified repair stations. If your mechanic is located at the same airport as one of these outfits, the tests could easily be combined.

Airworthiness directives

Airworthiness Directives (ADs) can be either one-time or recurring. One-time ADs typically specify that a particular component be inspected, modified, or replaced. The AD will specify how soon the operation must be performed; generally based on flight time, like, "Within the next fifty flight hours." Sometimes the time interval is sufficient to let you schedule the AD work in conjunction with the next annual.

Recurring ADs specify that a special inspection must be performed at given intervals. These range from fairly short (which can usually be canceled through the replacement of the offending component with an improved one) to several hundred hours. Similarly, these inspections may be relatively simple, or may require the use of special techniques such as dye penetrant and ultraviolet light. The existence of such ADs is one of the things you should have determined during your shopping phase.

Normally, the inspection interval is well-enough spaced that it can be combined with an annual. It's easy enough to predict if the aircraft's flight time will cross the magic number in the coming year.

Logbook entries

It's not enough to *do* the inspections, though. They must be entered properly in the appropriate logbooks. It's the responsibility of the mechanic to make the log entries. But it's in your best interest to ensure the entries are as precise and accurate as possible.

At a minimum, the person performing the inspection or procedure must include the type of procedure performed, the date, the mechanics' signature, and his or her certificate number. But, in most cases, more information is necessary. For the altimeter/static system inspection, the maximum altitude to which the test was performed should be included.

For preventative maintenance, enough details should be provided to ensure anyone reading the log 10 years hence will understand exactly what was done. "Rebuilt right brake" is insufficient. You'd prefer the log to read something like, "Honed right brake cylinder, replaced O-rings part #XXX-XXX and left and right brake pads, P/N XXXX-XXX. Replenished system with MIL-H-5606BB brake fluid." And again, the mechanic should date, sign, and include his or her certificate number. The gory details don't have to be included if there's a manufacturer's standard procedure to reference: "Rebuilt right brake in accordance with Service Bulletin 82-105."

For ADs, precision is even more important. Five years from now, you may be selling the aircraft. Some bright chap (who undoubtedly read Chap. 8 of this book) may look through your logs and say, "You didn't perform AD 93-10-4."

You'll point and say, "Here it is, 'Brake AD complied with'."

But of course, nothing says *that* particular AD was performed. At the minimum, the log entry should include the AD number and issuance date, the number of hours on the aircraft/component at the time of compliance, how the AD was complied with (i.e., inspection, replacement, and so forth—you can reference a section in the AD itself) and when the procedure will next be required. As ever, the entry must be wrapped up with the date of compliance and the signature/certificate number of the mechanic.

The FAA also requires that a number of other factors be tracked and recorded. They don't necessarily have to be kept in the aircraft logbooks, but must be permanently retained. You might as well put 'em in the logs as anywhere else.

14CFR 91.417 lists the items that must be recorded. In addition to the data on inspections, calibrations, and AD compliance, you must track:

1. The total time in service of the airframe, engine, and propeller.

2. The current status of life-limited parts of each airframe, engine, and propeller.

3. The time since last overhaul of all items installed on the aircraft, which are required to be overhauled on a specified time basis.

PROTECT THOSE LOGBOOKS!

From the preceding, it's obvious you'd be in a world of hurt should your airframe, engine, or propeller logs disappear. As mentioned in Chap. 5, a no-logs airplane is in a sticky situation. For instance, you'll have no proof of engine or airframe time nor records of compliance with any AD.

I kept my 150's logs in the aircraft; I'm smarter now. Contrary to what some folks say, the FAA does *not* require the logbooks to be carried aboard the airplane. You are required to present them only upon reasonable notice.

So find a safe place for them.

THE ANNUAL

To be legal to fly, your aircraft must have undergone an annual inspection in accordance with 14CFR Part 43 within the past 12 calendar months.

What constitutes an annual? Basically, the inspector is required to inspect the aircraft in order to determine whether it meets all applicable airworthiness requirements (Fig. 11-2). He or she is must use a checklist; either one of the inspector's own design or one provided by another source, like the aircraft manufacturer. Then

Fig. 11-2. *The annual is an in-depth analysis of the aircraft's airworthiness and its compliance with its Type Certificate. (Courtesy Mike Furlong.)*

the aircraft engine must be run to determine if its performance is satisfactory. The inspector is required to check the static and idle RPM, the magnetos, the fuel and oil pressure, and the oil and cylinder temperature.

Who performs it?

By 14CFR 65.95, only an A&P holding an IA is allowed to perform an annual on a standard-category airplane. What's the difference between an A&P and an A&P with an IA? The IA is required to have both the knowledge and the reference material to be able to determine whether a particular airplane still complies with its Type Certificate.

To gain an IA rating, an applicant must have held an A&P license for at least three years, must have been "actively engaged" in maintaining aircraft for at least the past two years, must have the equipment and facilities available to perform the tasks of an IA, and must pass a written examination. In addition to performing annual inspections, IA are authorized to inspect and return to service aircraft that have undergone major repairs or modifications and supervise "progressive inspections;" a special form of annual.

The IA rating expires on March 31st and must be renewed every year. At renewal time, the applicant must prove that he or she still meets the qualifications for gaining the license and has attended a refresher course in the previous year. In addition, he or she must have been actively performing the inspections, which require an IA rating and pass an oral exam by an FAA inspector on their knowledge of FAA Regulations and workmanship standards.

At some facilities, your aircraft will be inspected by an ordinary A&P. He or she will find the trouble spots and correct them. Then an IA will be called in, inspect the aircraft, and sign the logs.

If you've purchased an Experimental Amateur-Built aircraft (Fig. 11-3), the annual may be done by an A&P. No IA rating is needed, since Experimentals don't receive

Fig. 11-3. *Experimental Amateur-Built aircraft can receive their annual condition inspection from the original builder or any A&P—an IA isn't required. (Courtesy Ed Hicks/Van's Aircraft Inc.)*

Type Certificates. If your plane is a Special Light Sport Aircraft (the production-type LSA), then either an ordinary A&P or a person with a Light Sport Maintenance Repairman Certificate can perform the annual.

Selecting the shop

Where should you take your plane for the annual?

In the first place, the shop should be familiar with the type of aircraft you own. If you've bought a Cherokee or a 172, you shouldn't have any problems. With a Champ or Stinson, or other classic airplane (Fig. 11-4), though, you'll want to find an outfit familiar with the foibles of these older birds. It makes no sense to pay someone to learn how to work on your type of plane. There are a number of A&Ps who haven't done fabric repairs since their student days. Some mechanics, especially the smaller independents, may even refuse to work on certain types. "I won't touch the ------," one said to me. Apparently, the inspection panels for this model were all custom made—they only go on one spot, in one particular orientation.

In my opinion, the best solution is to get recommendations from fellow airplane owner. Join local pilot's groups or EAA chapters and ask around, you'll get a variety of opinions.

Of course, no one is perfect or unbiased. I was real happy with the man who did my 150's inspections. He was thorough, and recognized that I didn't have a lot of money for nonrequired repairs. Imagine my surprise, then, when I saw a note on a local airport's bulletin board warning *against* my mechanic. Didn't make any difference to me.

Make no mistake about it, picking the right mechanic can make a significant difference to your pocketbook. Like all businesses, there are a few who are less than ethical. Even worse, they can impart significant leverage that the aircraft owners can find hard to fight. If they feel certain repairs are required, they'll refuse to sign off your annual until the work is performed. And since your plane is lying half-disassembled on their shop floor, it's hard to take it to a different mechanic for even a quote.

Fig. 11-4. *This Luscombe Sedan is a rare bird; finding an experienced A&P/IA to perform its annual may be difficult.*

So chose carefully before turning your aircraft over. And remember, sometimes aircraft *do* require repairs. If the mechanic does find a problem, come in and take a look at it before starting to point fingers.

Cost

Talk over the anticipated cost prior to selecting the shop. Aircraft service manuals indicate the amount of time an annual should take; that should be the starting point. The service manual times vary by aircraft type. The Cessna 150 manual estimates 13 hours for an annual; that's actually more than mine took. A 172 takes about 16 hours, and a retractable might take 20 hours or more.

Typical shop rates are quite reasonable, really. I've had my airplane worked on for a lower per-hour rate than the local General Motors dealership charged. Still, the annual on that 172 might cost over $1100 with a typical $70 per hour shop rate. Some shops operate on a "flat rate" basis: The basic charge for the annual is based on the time predicted in the service manual multiplied by the shop's labor rate. This price includes all normal activities, plus, usually, most minor repairs. Parts are usually extra.

There are plusses and minuses to flat rate annuals. It does allow one to predict more closely what the annual will cost. Most problems found during annuals are pretty minor. These little problems can add up with a shop that charges by the actual time worked; it wouldn't change the price of a flat rate operation. There are obvious potentials for abuse if the annual charge is not capped somehow.

On the other hand, the shop performing a flat rate annual loses money if your plane takes longer than the shop manual's recommended time. There might be a tendency to cut corners; little problems may be ignored. Plus, make sure you understand exactly what is covered under the flat rate. The threat posed by an unscrupulous operator is obvious.

One way to reduce the cost of the annual is by participating in it. Many shops have reduced rates if the owner does the "grunt" work; wash the plane and engine, undo the cowling and inspection panels, remove the interior, and so forth. One flat rate shop I talked to said they reduced the time by 3 hours if the owner participated.

It's not for everyone, of course. A Bonanza annual might take three days to a week; that's a lot of time off for a $210 saving. If your time is shorter than your money, there's a lot to be said for simply turning the airplane over to the shop.

Should you decide to participate, find out the ground rules in advance. When will they actually need you? It might be only at the beginning and the end. If all you'll be doing is cleaning and reassembling the plane, then what tools should you bring? Most professional mechanics prefer not to have let others rifle through their toolboxes; you may be asked to bring your own screwdrivers and wrenches.

Finally, if there's other work you'd like to perform on the aircraft, the annual is the time to do it. After all, the plane has to be opened up anyway. Perhaps there have been some slight problems (Fig. 11-5), or maybe you'd like to have some changes made. Make sure you discuss it with the mechanic, and that you understand how much the additional work will cost.

Fig. 11-5. *The annual is a good time to get small glitches fixed. Here, an electronic thermometer and probe is being used to check out an oil-temperature problem.*

Scheduling

A nice thing about the annual is that it's predictable: You know when the previous annual will expire: 12 calendar months after the last one. If the annual was signed off on July 10, 2010, it's good through July 31st, 2011.

However, there is room for a bit of trickery. The applicable FAR (91.409) doesn't require a yearly annual; it requires that one be performed within the past year. But say you live in a colder climate where you won't fly through the winter. Assume that your new plane's annual is due in December.

There is nothing wrong with leaving your first annual until the spring thaw. Indeed, having a mechanic pore over your plane at the start of the flying season rather than at the end would probably be safer all around. You can also use the calendar year provision to gradually shift your annual date to one that is more convenient.

My Fly Baby's annual used to be due in August. But the summer months are the busiest flying time, especially for an open-cockpit airplane! I didn't want the plane laid up for a week during this period. So I started scheduling the annuals at the end of the month—timing the first one so that the mechanic's signoff was dated the first of September. That made it valid until the end of the following September. I repeated the process over a few more years, and now my annual comes due in March. This is perfect; the weather is warm enough to work in the unheated hangar, and my plane gets a good going-over prior to the start of the summer flying season.

Delivery

When delivering the aircraft to the shop, it's important to be as ready as possible. One thing the IA will check for is if all the required documentation is aboard the aircraft.

Make sure your airworthiness certificate, registration, radio license, weight and balance, and operations manual are there. They'll also need your airframe, engine, and propeller (if applicable) logbooks. Not just to make the final notation—they have to verify that ADs have been complied with and determine the total time in service of various components. They'll need contact information for you, in case something turns up. Finally, if you're having other work done, include a written list of squawks and problems. Be as specific as possible—tell exactly under what conditions the problems occurs.

Functions to be performed

As mentioned earlier, the IA must use a checklist to guide the inspection process. The checklist, no matter what its source, probably began with the guidance found in 14CFR Part 43, Appendix D.

It requires that all inspection plates, access doors, fairings, and cowlings be removed, and the engine and airframe thoroughly cleaned (Fig. 11-6). After that follows a whole smorgasbord of inspections: Fabric and Skin, Cabin and Cockpit, Engine and Nacelle, Landing Gear, Wing and Center Section, Propeller, Controls, and Radio (Fig. 11-7).

Of course, Part 43 breaks down the inspections to far greater detail. Here's what it says for just one section: "(d) Each person performing an annual or 100-hour

Fig. 11-6. *A thorough cleaning of both the engine and the airframe are part of any annual.*

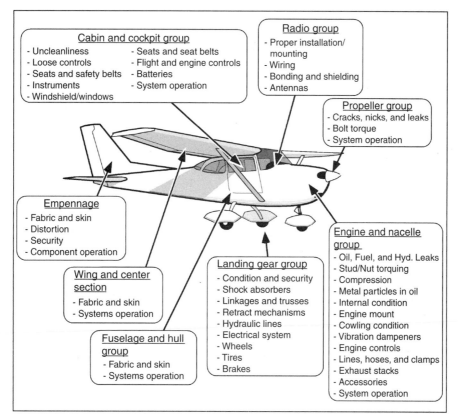

Fig. 11-7. *Annual Inspection target areas.*

inspection shall inspect (where applicable) components of the engine and nacelle group as follows:

1. Engine section–for visual evidence of excessive oil, fuel, or hydraulic leaks, and sources of such leaks.

2. Studs and nuts–for improper torquing and obvious defects.

3. Internal engine–for cylinder compression and for metal particles or foreign matter on screens and sump drain plugs. If there is weak cylinder compression, for improper internal tolerances.

4. Engine mount–for cracks, looseness of mounting, and looseness of engine to mount.

5. Flexible vibration dampeners–for poor condition and deterioration.

6. Engine controls–for defects, improper travel, and improper safetying.

7. Lines, hoses, and clamps–for leaks, improper condition and looseness.

8. Exhaust stacks–for cracks, defects, and improper attachment.

9. Accessories–for apparent defects in security of mounting.

10. All systems–for improper installation, poor general condition, defects, and insecure attachment.

11. Cowling–for cracks, and defects."

Some list, and this is just for the engine! In addition to the Part 43 tasks, the aircraft service manual identifies specific operations, which must be performed at each annual. It'll specify lubrication points, for example. And it may provide specific details for how to inspect certain parts.

The mechanic will either make a list of problem areas as he or she goes, or will fix each problem as it is discovered.

Completion

Eventually, it comes time to pay the (gulp) bill and pick up the aircraft. Check the logbooks and make sure the IA has made the required entry. It should be similar to the entries your instructor made in your pilot's log: "I certify that this aircraft has undergone an annual inspection on July 10, 2009, and is airworthy." The IA should sign it and include his certificate number. My mechanic has a sticker with the appropriate verbiage—he just sticks it to the appropriate page of the logbook and signs it. The same entry should be made *in all logbooks* (engine, airframe, and propeller). Check!

If any AD procedures were performed, the appropriate log entry should also have been made: "Oil Filter modifications made in accordance to Section 2 of AD-75-01." Completed with signature and certificate number, of course.

After settling up with the shop, don't hop in and fly just yet. Sad to say, even the best mechanics sometimes forget something. It's important to perform a thorough preflight after any sort of maintenance. Don't head right out on a cross-country, either. Stay in the pattern; do a couple of touch-and-goes. Land, taxi back to your tiedown spot, and check things over *again*.

Sigh, and wipe the sweat off your brow. It's over for another year.

REPLACEMENT PARTS

One thing you'll really cringe at is the price of replacement parts for your bird. I'm not going to try to justify the manufacturer's pricing schemes. There are some good reasons why aircraft parts cost more than the equivalent hardware-store items.

The Type Certificate

To prove the original design's safety, the manufacturer performed exhaustive structural analyses. Aircraft were ripped apart in ground test fixtures. Prototypes generated reams of flight test data. Once the FAA was convinced the aircraft met its standards, the design was granted a Type Certificate (TC). The TC actually defines the aircraft, describing the components down to the last bolt.

The TC isn't a mere bureaucratic ticket to be punched on the way to production. If the aircraft conforms to its TC, it automatically meets the same standards as the prototype. The TC is the pilot's guarantee: Barring improper maintenance, any particular aircraft will perform and handle like any other of the same model. Unless an aircraft conforms exactly to the components and construction shown on its TC, it cannot be legally flown.

Herein comes the rub. The TC specifies, which part numbers by which manufacturers are installed where. If you replace any part with one not specified in the TC, you've just busted the regs.

Legal replacements

What kind of parts *can* you install, then? Obviously, you can use parts built by the component's original manufacturer. Called Original Equipment Manufacture (OEM) parts, they are identical to what should already be installed in your aircraft. They can be obtained new through a parts supplier, or used from another aircraft.

Aircraft salvagers are good sources for parts, especially for those little trivial components that the manufacturers seem to delight in overcharging for. My old 150 had a pull-cable starter. The starter gear was engaged manually when the pilot pulled on a cockpit handle. A simple system: one not subject to the bendix gear woes of later models of the O-200.

Until both "ears" broke off the handle. I needed a new cable, and Cessna wanted almost $100 for the part! It's interesting to note that the local auto-parts store had a similar unit for $10. But it wasn't an approved part. And who wanted a handle labeled "choke" in their airplane? Instead, I called a few aircraft salvagers. One had the cable; it cost $30, plus shipping.

For components such as magnetos and starters, there is some paperwork involved. You'll hear the word, "tagged": yellow-tagged, green-tagged, and red-tagged. The term means that the used component's condition has been formally evaluated by a certified repair station. Yellow means the part is serviceable and airworthy. Green indicates the part isn't currently airworthy, but is repairable. Units in real bad shape earn red tags; fit only for scrap.

The parts you get from the salvager probably won't have a tag. In this case, you and your mechanic take the responsibility for using the component. As of this writing, it's legal. But bogus aviation parts have been causing problems lately. The FAA is considering a requirement that all parts installed on an aircraft must have a traceable heritage; that would essentially require that all installed parts have yellow tags. If you use a yellow-tagged part, retain the tag in the aircraft logbooks.

Parts for your aircraft *can* be made by someone other than the original manufacturer. Other companies can earn Parts Manufacturer Approval (PMA) from the FAA by proving that their parts meet the same requirements as the OEM parts. They can do this by making their parts identical to the OEM ones, buy gaining a license from the OEM manufacturer, or by performing the engineering required to prove the part is as good or better than the original. Such parts are marked "FAA-PMA" along with information regarding where the part is authorized for use.

Owner-manufactured parts

There's another source of parts for your airplane—make them yourself. 14CFR 21.303 allows the owner of an aircraft to manufacture parts for that aircraft. The important point is that the part manufactured must have all the characteristics of the original, FAA-approved part (Fig. 11-8).

Fig. 11-8. *As the owner of an aircraft, you are authorized to build parts for it, such as cut, bend, drill, and rivet replacement structures. However, the parts must conform to the original Type Certificate—you can't change the design.*

For example, imagine a slight mishap damages a piece of aluminum skin on your aircraft. The aluminum is alloy 2024T-3, 0.025-inches thick. You can buy a sheet of the same aluminum alloy, cut the part out, drill the rivet holes, and your A&P can install it on the aircraft and legally sign it off.

The key point is that the part must have all the characteristics of the original part. If it's a steel part, the material must be heat-treated to the same extent. If it bolts on, the same size bolts must be used. In the example above, you must use aluminum of the same alloy, temper, and thickness as was originally installed on the airplane. You don't actually have to hammer out the parts yourself, but the FAA requires that the owner be closely involved with their manufacture. The owner has to:

- Provide the manufacturer of the part with the design or performance data, or
- Provide the manufacturer of the part with the materials, or
- Provide the manufacturer with fabrication processes or assembly methods, or
- Provide the manufacturer of the part with quality control procedures, or
- Personally supervise the manufacture of the new part

For any part—OEM, STC, or owner-produced—to be legal requires that the part be properly designed, conforms to that design, production is properly documented, and the part is properly maintained once it's in service.

Standard hardware

Another exception to the OEM/PMA parts requirement deals with standard hardware. The US Government, over the years, has developed a number of standards for the construction and testing of aircraft-grade hardware. The aircraft and component manufacturers will specify nuts, bolts, washers, and other parts based on one of the following specifications:

1. MS (Military Standard)
2. NAS (National Aerospace Standard)
3. AN (Air Force/Navy)

These specifications will list the minimum, and, in some cases, the maximum requirements for approval. These requirements include dimensions, tolerances, strengths, and finishes. Most of the hardware used on your aircraft will probably be of the AN variety. This is the granddaddy of standards; in fact, the AN used to stand for Army-Navy.

"What's so special about the AN standard? Why can't I just buy bolts at the hardware store?" For one thing, the ordinary bolts from the hardware store aren't very strong. The most common units are about a third the strength as the same-size AN bolts. Second, AN specifications usually require that parts be cadmium-plated. When steel comes in direct contact with aluminum, a little moisture can trigger a phenomenon called *dissimilar metals corrosion*. The cadmium plating of an AN bolt prevents its steel from coming in direct contact with whatever it's holding together. Finally, the most important thing about the AN standards is that they are just that, *standards*. The bolt maker certifies that every unit meets the requirements of the specification. Maybe the bolt in the hardware store bin this week is just as strong. But the one lying in the same spot next week might not be.

Your mechanic will replace hardware in your plane with parts meeting AN/NAS/MS standard—make sure you do the same.

MODIFICATIONS

There isn't a pilot in the world completely satisfied with his or her airplane. Some deficiencies are fondly regarded as quirks; mere beauty marks on their beloved's cheek. Serious problems usually result in a change of ownership. Between these two extremes, the decision is harder. Sure, you like the plane, except, well, it doesn't have good short field performance (Fig. 11-9), or it won't fly far enough, or the FAA says you have to install a new black box.

But the government is real persnickety about aircraft modifications. The FAA is perfectly willing to approve almost any change that won't effect the reliability or safety of the aircraft. However, the burden of proof is on the modifier to prove that the altered aircraft meets the same safety requirements as the original design.

We've already covered the process the manufacturer went through to get a Type Certificate for your aircraft. And how, if your plane doesn't conform exactly to the components and construction shown on its TC, its airworthiness certificate is invalid.

However, the situation isn't totally rigid. Exceptions are listed in the aircraft's minimum equipment list. The FAA also allows minor modifications under the field approval process, using the FAA Form 337. A wide range of changes may be made, but the alterations cannot affect the aircraft's performance, handling, or flight characteristics. If they do, (or, in the opinion of the FAA, there is a chance they might) the mod must be thoroughly analyzed to support the issuance of an addendum to the original TC: A Supplemental Type Certificate, or STC.

Let's examine Form 337 mods first.

Fig. 11-9. *A 65-HP Standard Chief (right) meets a 115-HP Super Chief. The speed increase of this STC'd modification pales in comparison to the plane's new short-field ability.*

Field-approval modifications

The most common field approval is for avionics installation. The Form 337 can be executed by any A&P with an IA authorization. The FAA Flight Standards Maintenance Branch must eventually endorse the completed paperwork, but the aircraft can be returned to service once the IA signs the form. That way, you aren't grounded waiting approval for a routine radio installation.

However, Flight Standards is not a rubber-stamp organization. If they feel performance, handling, or flight characteristics will be affected, field approval will be denied. If a change is unusual or the IA has any doubts, he can refuse to return the aircraft to service until the FAA has actually signed the paperwork.

Your best move is to obtain Flight Standard's approval before the modification is made. One big difference between a field approval and an STC is that the STC process may require inspection and documentation of the work in progress. If you've already made the changes, and the FAA decides an STC is required, you'll have to open the aircraft up, enough to allow the internal parts to be inspected.

A week's work of research and discussion before the change started can avoid months of argument and downtime. Flight Standard's Airworthiness Inspectors are reasonable. If you can find a similar mod performed under a Form 337, they'll probably sign off yours. One pilot bought a Cessna 195 in Canada and brought it into the States. However, it had a different engine installation, and his local Flight Standards office refused to grant a field approval. The owner talked with inspectors at another region's Flight Standards offices, and found one who had approved the same configuration. After a short three-way conversation, the local office approved the change.

Field approval is a relatively inexpensive process. Other than the actual costs to alter the aircraft, the main expense is the IA's time to recalculate weight and balance and to describe the modification on the back of the Form 337. Laying careful groundwork with Flight Standards beforehand will smooth FAA approval. However, if they won't accept the change as a field modification, an STC is the only option. You're facing some pretty involved engineering analyses and structural tests. You'll probably need some hired guns, and they're not cheap.

Buying an existing STC

Before starting a range war with the FAA, perhaps there's an easier way. Is the same modification sold in the form of an approved multiple STC? The FAA maintains a summary of available STCs. It supplies a brief description of each modification, and the address of the STC holder. The hard-cover version is mighty big document—two thick volumes of fine print. The section on Cessna 210s alone is more than 11 pages.

Most owners find out about available modifications through advertisements. However, there are a few things to watch for. First, the modification must be FAA approved under a multiple STC. Beware of the terms "fits," or "adaptable to," for instance. Just because something can be bolted onto an aircraft doesn't mean the FAA has okayed it.

If the STC holder sells components to support the mod, the company must possess a Parts Manufacturing Approval, or PMA. Otherwise, the parts aren't legal. An interesting quirk in the regulations allows the STC holder to modify aircraft in his own facility without the need for a PMA. No coordination with the FAA is required when performing a change in accordance with a purchased STC. The IA signs off the work and notifies the FAA via the same Form 337 used for field approvals. That's it.

The cost of buying rights to an STC varies. Most older aircraft were certified to use cotton fabric, so suppliers of more modern covering materials practically give their STCs away. If they didn't, most customers couldn't legally use their products. On a slightly higher scale, the EAA charges a buck a horsepower for their auto-fuel STC. The rights for some major mods can run in the thousands of dollars, not including labor or materials. It may seem like a lot of money, but not compared to the difficulty of earning your own STC.

Do-it-yourself STC

If an appropriate mod isn't on the market, you may have to plunge ahead and gain your own STC. To begin with, any part installed on a certified aircraft must be of aircraft quality. This can be proven in two ways: Either the part has been manufactured under Technical Standard Order (TSO) provisions, or the modifier must prove that the part is qualified under the airworthiness FARs (Fig. 11-10).

To legally install a non-TSO'd part like an automobile alternator, you must prove that the alternator meets the same requirements as the aircraft unit. If the TSO requires the alternator generate full-rated output at 150°F while subjected to a 3-G vibration, you must present test data to that effect.

"But the parts are identical!"

Fig. 11-10. *To STC a track installation on your Cessna, you'll need to prove that the tracks meet the provisions of the original gear-and-wheel TSO.*

Are they really? Sure, the cases may match. But are the field windings the same wire gauge? Does the auto unit use the same bearings as the aircraft unit? Is the built-in regulator an identical design, with identically-rated components? It doesn't make much sense to risk thousands of dollars in avionics just to save a hundred bucks on an alternator. Or risk even more, if the alternator fails during IFR or at night.

Just because the part has TSO approval doesn't mean you can install it. The TSO is essentially just a bench check; it's up to you to prove that it functions properly when installed in a particular aircraft. Once you've picked out the TSO parts to be used, the first stop is at the FAA's regional Aircraft Certification Office (ACO) to explain the proposed modification. The ACO is the approval agent for STCs. Again, talk to the FAA before you start work. The engineers of the ACO are only too happy to discuss your proposal. They'll explain the regulations and assist you during the certification process, but they aren't allowed to help design or substantiate a modification.

The ACO first determines the "certification basis" of the aircraft from the original type certificate. For the modification to receive approval, the aircraft must still meet the minimum requirements in effect at the time of its original certification. For instance, the stall speed of a FAR-23-certified aviation aircraft is limited to 61 knots in the landing configuration. Clip the wings as you will, but an STC cannot be granted unless the stall speed remains 61 knots or less. An older aircraft may have been certified prior to this rule, and needn't comply. However, the older requirements were occasionally more strict. Planes like the Aeronca Champ must recover from a six-turn spin hands-off, while modern normal-category aircraft need to only stop spinning within one turn or 3 seconds of the application of normal recovery techniques.

Exemptions can be made. An example would be a turbine engine installed on an aircraft built before turbines were certified. The ACO would use more modern regulations where the engine is concerned. Compromises between older and newer regulations are allowed, but it's up to the ACO.

The regulations specified in the certification basis identify those factors, which must be proven by engineering analysis or ground testing versus those, which require flight testing. The ACO will determine the amount and complexity of the required analysis and tests. Some leeway is allowed. If the proposed change is minor, the ACO may waive the need for extensive analysis or testing.

The basic objective of the engineering analysis is to prove that the forces and load limits will remain within the allowable range. For example, in a case where a larger engine is to be installed, the problem isn't restricted to the engine mount. Torque, gyroscopic forces, and slipstream effects, among others, must be considered. The heavier engine may make a beefier landing gear necessary. What about the CG? A shorter engine mount might be required to keep the CG in the proper range, and/or some equipment shifted aft. Sometimes, excessive loads result in a category change from aerobatic to utility, or utility to normal.

The FAA doesn't care who does the analysis, but most of us lack the right background. Any knowledgeable engineer would be adequate, but the best way is to hire a Designated Engineering Representative (DER). Not only can DERs perform the analysis, but they can officially make decisions on behalf of the FAA with regard to whether a particular modification complies with regulations. To qualify as a

DER, a candidate must have eight years of aviation engineering experience, plus one year working in aircraft certification with the FAA.

A DER is used to working with the FAA, and understands its requirements. In turn, the ACO engineers are familiar with the DER's work. The annual renewal of the DER's appointment, reviews of portions of his or her submittals, and professional pride ensure thorough and precise output. Performance of the work yourself requires an exhaustive review of your submittal by the ACO engineers. This is understandable, since they must decide whether the modification complies with the FARs. Since a DER is authorized to make these decisions, approvals are gained far faster.

On the minus side, the DER is a private citizen and charges for his or her services. The going rate is $100 an hour or more; the smallest mods will probably take an absolute minimum of at least 10 hours. DERs are appointed for particular specialty areas, such as flutter or structure. Involved modifications may require hiring more than one DER.

Whoever does the analysis will need to know the strength of the aircraft's structure at various points. But the original manufacturer has no obligation to provide this data, and has considerable financial and legal incentive not to. The DER might be forced to "reverse engineer" the design, further adding to the expense of the STC.

The fact that another modifier has gained approval for the same modification doesn't help. Supplemental Type Certificate data is proprietary information; it cannot be used in the approval of another STC unless the owner grants permission. This differs from the field approval, where a single precedent can open the flood gates. Each STC application is evaluated solely on the data submitted.

After all, fair is fair. The EAA spent several years and considerable money to gain their auto-fuel STC. If the FAA had allowed free use of proprietary data, other companies could use the EAA's data to market their own STCs. The EAA would have no chance to recoup its expenses. If this were the case, why would anyone go through the difficulty of gaining approval and marketing an STC?

As mentioned, the purpose of the analysis is to verify compliance with applicable certification requirements. These often demand physical proof as well. The ACO will identify both ground and flight tests, which must be performed. Ground tests can be simple, such as a flow test from a newly-installed fuel tank, or quite involved. Trying to certify a taildragger conversion of a trigeared aircraft (Fig. 11-11)? Plan on loading the plane to gross, then hoisting it up and dropping it from various heights to prove the strength of the new landing gear. These tests are dangerous; it is quite possible to overstress the airframe to the point where it is no longer airworthy.

If a modification requires flight testing, the aircraft usually needs temporary certification in the experimental/research and development category. The FAA will inspect the aircraft throughout the modification process. Upon completion, the Experimental Certificate of Airworthiness is issued and flight testing can begin. While these inspections and the certification are free, the ACO's workload may cause delays. If time is a factor, Designated Airworthiness Examiners (DARs) can perform the inspections and issue the certificate. But just like DERs, the DARs are private citizens who charge around $50 an hour.

Anyone with a private license can play test pilot. The flight tests verify those details, which haven't yet succumbed to the engineer's art, such as engine cooling

Fig. 11-11. *If you want to develop your own taildragger modification, you'll be expected to perform extensive testing. It's easier to just buy a commercial STC package.*

and handling qualities. The experimental/research and development certificate may require occupant parachutes and the installation of jettisonable doors, especially for spin or flutter testing. Once the flight test program is completed, an FAA test pilot will verify the results.

At this point, it's all over, but the paperwork. In addition to the engineering analysis and flight test results, the modification must be documented. In the case of one-time STCs, this can consist of detailed sketches (Fig. 11-12) and photographs. However, if trying for a multiple STC, drawings and diagrams sufficient to duplicate the modification must be submitted.

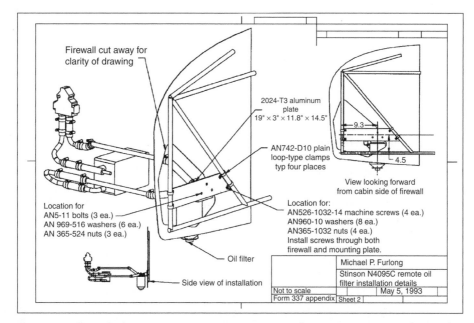

Fig. 11-12. *Each STC application must be accompanied by detailed drawings. They can be done by hand, but must be clear and descriptive. Michael P. Furlong.*

Once the FAA signs off, it's done. The aircraft has no restrictions beyond those of its certification category. There's one gleaming star in this whole mess: there's absolutely no charge by the FAA. Small consolation while you're writing out checks to the A&P, IA, DER, and DAR.

Is this trip really necessary?

Sounds arduous, doesn't it? Before you seriously consider modifying your airplane, you'd better make sure you'll be happy with it.

Aircraft design is an exercise in compromise. A long-range airplane should carry lots of fuel; yet the extra weight reduces short-field capability. Airfoils which maximize cruise speed often have vicious stall characteristics. Exotic drag-cheating shapes increase manufacturing expense and the price charged for the completed airplane.

Modifying an aircraft is like poking your finger into a tightly-stacked pile of bricks—one brick may slide back, but another gets shoved out somewhere. The modification process merely adjusts the compromises inherent in the design. It's too easy to focus on the "plusses" and neglect the drawbacks. For example, examine the gains and losses of installing a larger engine. Cruise speed will increase (although not much as you might expect), as will the climb rate and the ability to get out of short fields. However, all the airframe-related limitations remain—you might be able to cruise faster, but red-line and maneuvering speeds restrict how much of the capability can be legally used. Fuel consumption increases and range gets shorter.

Compensate by adding long-range fuel tanks? This corrects the range problem, but another problem arises. Since the legal gross weight of the aircraft doesn't change, the weight of the heavier engine and extra fuel reduces the useful load. A four-seat airplane becomes a three seater.

There are also subtle effects which one might not expect. A heavier engine brings the CG forward; a common correction is to move the battery aft to compensate. However, this affects the inertial properties of the aircraft—once in a spin, it has a greater tendency to keep spinning.

When buying the rights to a multiple STC, get the performance figures beforehand. Do a little mental arithmetic. Assume you're buying a brand new airplane. Does it still meet your needs? If planning your own STC, talk to an experienced IA or the engineers at the FAA Aircraft Certification Office. They should be able to give some indication of the drawbacks of the proposed modification.

Executing an STC is a lot like making a parachute jump. Its very difficult to get back into the same airplane if you find out you were wrong. Don't let the "plusses" blind you to the "minuses."

Summary

As far as the FAA is concerned, aircraft aren't modified in workshops or hangars. To them, it happens in offices, where the aircraft's type certificate is modified or logbook entries reflecting Form 337 mods are made. The FAA's overall concern is safety, which often has more to do with analysis and test results than with aluminum and steel.

Of course, the person making the modification must incorporate all factors. The paperwork jungle must be plowed; aluminum and steel must be worked. Advance coordination is the key, talk to the FAA before you start.

CASE STUDY: REWORKING A SKYLANE

While all Cessna Skylane buyers would prefer a shiny example fresh from the factory floor, the price tag for a brand-new bird is daunting. Javier Henderson took an alternate approach: He bought a 30-year-old Skylane with a tired engine, bad paint, and outdated avionics and turned it into a stylish bird practically identical to the new-production aircraft.

The California computer consultant had been a partner in an American General Tiger, but with his two kids getting older and bigger, it was time for a new airplane. "They were outgrowing the Tiger in room, range, payload, and CG," says Henderson. He and his wife settled on a 1973 Cessna 182. "We don't take a lot of long trips, so cruising at 135 knots was fine. We decided on a simple-to-maintain and relatively economical-to-own airplane."

Finding the Skylane was simple—it was based at his home airport, and his A&P told him about it. A previous owner had replaced the original avionics, but the new radios were obsolete by the time Javier looked at the plane. The engine was on its last legs, so the Hendersons took the replacement cost into account as when they made an offer. The purchase price was $69,000, which came from selling their Tiger ownership share.

The first year of ownership featured Javier putting Cessna 772CG into near-new condition. "I didn't do it all at once," he says. "It was done piecemeal so I could fly the plane and let the wallet recover."

Out went the outdated avionics, to be replaced by a new Garmin stack, including a GNS430 nav-com, GTX330 transponder, and MX20 multifunction display. A King KX155 nav-com is in the #2 slot and a S-Tec 55x autopilot was installed. All the engine gauges were replaced with Electronics International digital gauges, and a GEM 610 engine monitor was added (Fig. 11-13).

The old panel was scrapped literally! "The royalite overlay is now resting in some landfill," says Henderson. "The aluminum subpanels have been recycled into soda cans. In their place are solid, single-piece aluminum panels, powder-coated with the same paint that Cessna is using on Skylane panels these days."

The tired O-470 was replaced with a brand-new engine from LyCon Aircraft Engines in Visalia, California. It cost $22,000. A complete repainting cost another $12,000. By the time new paint went on, the Henderson's Cessna had completed its year of metamorphosis (Fig. 11-14).

Ownership experience

Javier and his family put about 125 hours per year on Cessna Two Charlie Golf. Some of the time is business-related, but the remainder is semilong family trips (generally 300 miles or so), local sightseeing, "$100 Hamburger" runs, and so on.

Fig. 11-13. *Javier Henderson upgraded the panel of his Cessna 182 with electronic engine instruments and digital avionics.*

He pays $95 a month for a closed hangar just ten minutes from home. Insurance, including liability and in-flight hull coverage, runs $1,100 a year.

He's been very pleased with the aircraft's reliability. "Given that all the radios and the engine were replaced, there's very little left on the airplane with any appreciable time on it." He's only had two cases of unscheduled maintenance in his three years of ownership—a leaky fuel-tank gasket and a bad microswitch on the flap motor assembly. The repair bill totaled $1,400. The on-off switch on his Garmin Nav/Com failed, and an AD mandated a software change on the Garmin transponder, but the company fixed both these items under warranty.

Fig. 11-14. *The crowning touch of the Cessna's upgrade was a new paint job.*

His first annual inspection was performed by the same shop that had handled his Tiger, but the last two have been owner-assisted with an independent A&P. Henderson says, "I did all of the work . . . removed the interior and inspection panels, did the maintenance tasks in the service manual, serviced the spark plugs, changed the oil, etc. The A&P/IA poked and prodded, checked compression and mag timing, etc." He replaced the aircraft battery on the first of the owner-assisted annuals, and a bad mixture cable on the second. The battery replacement is allowed owner-maintenance, and the A&P supervised and signed-off his mixture-cable work.

The annual at the full-service shop cost about $2,000, and he sees considerable savings with the owner-assisted versions. Generally, each annual requires about $500 in various parts. When he changes the oil, he sends a sample in for analysis— a practice, which helps him keep track of what's going on inside his new engine.

A summary of his ownership costs can be found in Fig. 11-15.

Advice

As far as selecting the plane is concerned, Javier Henderson recommends extensive research prior to shopping. "Since 182s have been in existence since 1956, deciding you want a 'nice Skylane' is just the beginning," he says. "The model has seen a lot of changes over the years, some quite profound."

He also recommends spending a lot of time researching ownership itself. "Be familiar with what the FARs let you work on, and plan on doing as many of those maintenance items as you can. Even if you need instruction on how to hold a screwdriver."

He subscribes to Light Plane Maintenance magazine (www.lightplane-maintenance.com). "I read it cover to cover every month, even if the subject doesn't pertain to my airplane. I may not perform every maintenance task they discuss, but it takes away some of the mystery when I have to take the plane to a shop."

"Regardless of how much you like to get your hands dirty," says Henderson, "be an informed owner. The worst thing you can do is hand the plane, keys, and logbooks to a mechanic and say, 'fix it.'"

Cessna 182
Owner: Javier Henderson

Hangar rent: $95/mo   Typical annual inspection: $500
Insurance (full): $1100/year   Other yearly maintenance: $500
Hours/year: 125

Fig. 11-15. *A summary of Henderson's ownership costs.*

12

Owner Maintenance

If you've ever done any maintenance on the family car, those skills can help achieve a significant drop in airplane ownership costs.

In some ways, aircraft are simpler to work on. Changing eight plugs on a '79 V-8 Monza calls for more contortions than doing the same job on a '47 Champ. As Fig. 12-1 shows, maintenance is a lot easier when you can remove a simple cowling and lay the engine bare in a matter of moments.

Owner maintenance isn't for everybody, though. It does take time. Plus, some folks just can't pick up a tool without sticking it through something. I once got fumble-fingered and ended up dropping half a gas cap into a fuel tank. Of someone else's airplane, no less.

Working on the plane is fun, but sometimes I'd like to just hand my wallet to a mechanic and eliminate the hassle. Unfortunately, my billfold is rarely thick enough to give my fumble-fingers any rest. This chapter is for those in the same boat. Here, we'll discuss what the FAA allows the owner to do, what tools are necessary, and some guidelines for aircraft-quality work. Then we'll take a look at various aircraft systems and owner-maintenance procedures associated with them. Finally, we'll examine some of the most common owner-maintenance procedures.

Before we get started, two points should be emphasized. First, there's no shame at being reluctant to perform some owner-maintenance tasks. Working on your own plane can enhance your confidence. But if you're skittish about a particular procedure, you're not likely to be comfortable while flying the plane afterward. By all means, leave it for the A&P.

Second, don't imagine you have to do it alone. Nervous about oil changes? Schedule your first one with the A&P, and have him or her show you how. Similar help is available through type clubs or local owner's groups. Even if you don't plan on doing any of your own work, at least skim through this chapter. You'll probably pick up some information that'll make it easier to converse with your mechanic.

Fig. 12-1. *Getting access to an aircraft's engine is a lot easier than most cars, but it's usually less effort when two people handle the cowling.*

BASICS

To begin with, let's examine some of the basics involved in owner maintenance.

Legalities

The authority for an owner to work on his or her aircraft is contained in 14CFR 43.3 (g): "The holder of a pilot certificate issued under Part 61 may perform preventative maintenance on any aircraft owned or operated by that pilot that is not used under Part 121, 127, 129, or 135."

The term *preventative maintenance* is of interest. The FAA defines it in Appendix A section (c) of Part 43. Those sections applicable to normal, utility, and acrobatic aircraft operated under Part 91 reads as follows:

"(c) *Preventive maintenance.* Preventive maintenance is limited to the following work, provided it does not involve complex assembly operations:

1. Removal, installation, and repair of landing gear tires.
2. Replacing elastic shock absorber cords on landing gear.
3. Servicing landing gear shock struts by adding oil, air, or both.
4. Servicing landing gear wheel bearings, such as cleaning and greasing.
5. Replacing defective safety wiring or cotter keys.
6. Lubrication not requiring disassembly other than removal of nonstructural items such as cover plates, cowlings, and fairings.

7. Making simple fabric patches not requiring rib stitching or the removal of structural parts or control surfaces. In the case of balloons, the making of small fabric repairs to envelopes (as defined in, and in accordance with, the balloon manufacturers' instructions) not requiring load tape repair or replacement.

8. Replenishing hydraulic fluid in the hydraulic reservoir.

9. Refinishing decorative coating of fuselage, balloon baskets, wings tail group surfaces (excluding balanced control surfaces), fairings, cowlings, landing gear, cabin, or cockpit interior when removal or disassembly of any primary structure or operating system is not required.

10. Applying preservative or protective material to components where no disassembly of any primary structure or operating system is involved and where such coating is not prohibited or is not contrary to good practices.

11. Repairing upholstery and decorative furnishings of the cabin, cockpit, or balloon basket interior when the repairing does not require disassembly of any primary structure or operating system or interfere with an operating system or affect the primary structure of the aircraft.

12. Making small simple repairs to fairings, nonstructural cover plates, cowlings, and small patches and reinforcements not changing the contour so as to interfere with proper air flow.

13. Replacing side windows where that work does not interfere with the structure or any operating system such as controls, electrical equipment, etc.

14. Replacing safety belts.

15. Replacing seats or seat parts with replacement parts approved for the aircraft, not involving disassembly of any primary structure or operating system.

16. Trouble shooting and repairing broken circuits in landing light wiring circuits.

17. Replacing bulbs, reflectors, and lenses of position and landing lights.

18. Replacing wheels and skis where no weight and balance computation is involved.

19. Replacing any cowling not requiring removal of the propeller or disconnection of flight controls.

20. Replacing or cleaning spark plugs and setting of spark plug gap clearance.

21. Replacing any hose connection except hydraulic connections.

22. Replacing prefabricated fuel lines.

23. Cleaning or replacing fuel and oil strainers or filter elements.

24. Replacing and servicing batteries.

25. Cleaning of balloon burner pilot and main nozzles in accordance with the balloon manufacturer's instructions.

26. Replacement or adjustment of nonstructural standard fasteners incidental to operations.

27. The interchange of balloon baskets and burners on envelopes when the basket or burner is designated as interchangeable in the balloon type certificate data and the baskets and burners are specifically designed for quick removal and installation.

28. The installations of anti-misfueling devices to reduce the diameter of fuel tank filler openings provided the specific device has been made a part of the aircraft type certificate data by the aircraft manufacturer, the aircraft manufacturer has provided FAA-approved instructions for installation of the specific device, and installation does not involve the disassembly of the existing tank filler opening.

29. Removing, checking, and replacing magnetic chip detectors."

Interpretation

Interpretation, as ever, is the key. Take the first half of the first sentence: "Preventive maintenance is limited to the following work. . . ." In other words, if the procedure isn't listed in this section, the owner *cannot* legally perform it. Simple enough—apparently. But take a gander through the list and point out where it authorizes the owner to perform oil changes.

It doesn't. It lets you clean and replace oil strainers. But nowhere does it authorize draining engine oil.

Sure, there's the "lubrication" clause. But my dictionary defines *lubrication* as adding oil, not draining it. Compare how subsections 16 and 17 spell out exactly what the owner can do with the position and landing lights—a far less critical system than the oil tank!

Fortunately, the FAA and the rest of the known world interprets this section to allow owners to change their own oil. Every time the need for maintenance arises, the owner must make a decision as to whether he or she is legally authorized to perform the procedure.

It's not cut and dried. An owner is authorized to remove and replace tires, for instance. Nothing in Appendix A allows the owner to remove and install the *tubes* inside those tires. Yet once the tires are off, the tubes are trivial.

The second half of the first sentence in Section (c) isn't of much help: ". . . provided it does not involve complex assembly operations." What's complex? Again, we're at the point where interpretation is up to the aircraft owner.

Basically, the FAA doesn't want you making an unwitting mistake that might threaten safety. If you have any questions about the legality of a particular procedure, contact the nearest Flight Standards District Office (FSDO).

However, owners are allowed to perform whatever operations on an aircraft they desire, as long as a licensed mechanic is willing to supervise and sign off the work in the aircraft logbooks. As one might expect, this is the sort of thing to arrange before taking a wrench to your airplane. But once your mechanic has a feel for your shop skills, you may be able to perform more advanced work. For example, I had a spate of solenoid problems in my old 150. After the third failure, my mechanic let me install the new one at my convenience and just bring it by for an inspection and signoff. There wasn't much to it, just four bolts and a couple of battery cables. He didn't charge me, either.

The FAA restrictions aren't there to generate business for A&Ps. They're to make sure a repaired airplane is *airworthy*, not just reassembled.

Finally, the final source for proper procedures and materials is the official service manual. Engine and airframe manuals are available through the manufacturer; for long-gone aircraft, other companies (such as Univair) republish them. If you're serious about working on your plane, get the appropriate service manuals.

Tools

You'll be needing some tools to assist in your aircraft work:

Screwdriver set. Both standard and Phillips are used. Get an assortment; you'll probably use them all, eventually. Power screwdrivers are jim-dandy, especially for removing all the inspection plates at annual time. Careful, though, especially while tightening. I once wrenched off a nut plate with an overly-enthusiastic electric screwdriver.

Socket set. U.S. measurement, 3/8-inch drive with ratchet wrench. A decent set will come with various adapters and extensions. A 1/2-inch drive set might be necessary for some jobs. A 1/4-inch drive outfit is a bit easier to handle for light work.

Torque wrench, 3/8" drive. You might be able to work by guess and by golly on your Ford, but don't try it on an airplane. There are two kinds of torque wrenches. One has a gauge readout of torque applied. On the other, you set the desired torque and the wrench "clicks" when you reach it. The second kind is easier to use.

A deep 7/8" socket. That's a *deep* socket; it'll be used to remove and replace spark plugs, and aviation plugs are longer than car ones. To be sure, take an old aircraft plug to the tool store when you go (your mechanic probably has some unairworthy ones lying about).

A set of combination, open-end and box-end wrenches.

Safety-wire twisters. These look like largish wire cutters, but have a central spiral shaft (Fig. 12-2). You usually have to buy these from an aircraft supplier. Price varies: $25–$75 new. I found a pair at a surplus store for $10.

Spools of 0.032 inch and 0.040 inch stainless steel safety wire.

A grease gun and a cartridge of the appropriate grease (check your service manual).

Spare screws and other small aviation hardware.

Pliers, vice grips, and other common hand tools.

Most of these can be bought at the neighborhood hardware store. Over the long run, you'll be happier with good-quality tools. In fact, aircraft ownership is a good excuse to buy good-quality tools.

It's most convenient to keep these at the airport in your locker; however, you may be reluctant to leave expensive tools unattended. When my plane was kept in

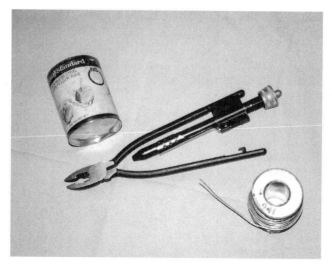

Fig. 12-2. *Safety-wire pliers are a required item for most owner maintenance.*

an open hangar, I kept a small tackle box of cheap screwdrivers and whatnot in my car (Fig. 12-3), and brought the good toolbox when there was real work to be done.

Holding it together

There are a lot of similarities between working on your car and working on your airplane. You'll use the same tools, and, for the most part, the procedures are similar.

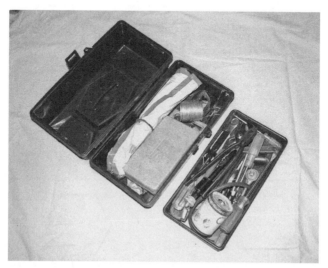

Fig. 12-3. *Portable tool kit for typical light maintenance. It contains a small socket set, a pouch of wrenches, some screwdrivers, the safety-wire pliers, plus wire, tape, and extra washers and nuts.*

However, there's one big difference between car and aircraft work: Aircraft must *not* come apart on their own. This is generally is the start—if not the end—of a very bad day.

Yes, you don't like your car to come apart on its own, either. But the aviation world is a lot more stringent. Something *must* provide assurance that vibration, heat, or stress will not cause components to separate. When we bolt something together around the home, we usually use just a nut and a bolt. We tighten the nut thoroughly to prevent it from coming loose. That's not good enough on an airplane. Take a close look. Have you ever seen so many self-locking nuts in your life? And cotter pins, and safety wire.

Aviation hardware is designed to both work, and *last*.

BOLTS

In the last chapter, we discussed aviation-quality hardware standards. Most bolts on your aircraft will meet AN specifications; Generally speaking, you'll find AN3- through AN20-type bolts. AN bolts always have markings on the head to indicate that they are high-strength bolts. A typical marking is a four- or six-pointed star, possibly accompanied by a couple of letters. Typical head markings are shown in Fig. 12-4.

If a bolt doesn't have markings, it isn't an approved bolt and shouldn't be used. However, lines on the bolt head is an industry-standard method of indicating high-strength bolts; the markings aren't exclusive to aircraft hardware. The only way to be sure is to buy the bolts through an aviation supplier.

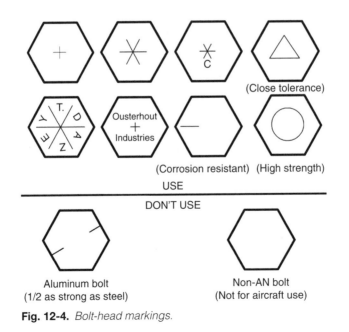

Fig. 12-4. *Bolt-head markings.*

The characteristics of a particular bolt is described by its specification. AN3-14A is a typical one:

- AN means the bolt meets Air Force-Navy standards.
- Three is the diameter of the bolt in one-sixteenths of an inch.
- 14 is a code for the approximate shank length.
- A indicates that the thread area isn't drilled for a cotter pin. You won't be able to use it with a castle nut.

Figure 12-5 shows how to decode bolt specs. There are other codes; for instance, a bolt with a head drilled for applying safety wire will have an "H" after the diameter code; "AN3H-10A," for instance.

The specifications for AN bolts allow some variance for diameter—some bolts might fit tighter than others. When aircraft designers want more precision, they'll specify a "close-tolerance" bolt. These are generally of the AN173 through AN186 series, and feature a triangle on the head. Similarly, sometimes a stronger bolt is required and there isn't enough room for a larger-diameter one. In this case, the designer may specify NAS1103-1112 series. These bolts feature a cup-shaped depression in the head.

NUTS

The nuts, naturally, are designed to work with the bolts. Most of the "secure" onus is placed on the nut; hence, aviation has a variety of locking or lockable nuts.

The AN365 elastic stop nut is an example. It has a semisoft plastic or fiber insert that grips the bolt's threads to keep the nut from turning. Look into the end of the nut to see the colored insert. This insert can cause some problems. Every time the nut goes on or off, it wears just a tiny bit. Eventually, it loses its self-locking ability. If you can tighten an AN365 stop nut by hand, discard it. Also, the insert is certified to only 250°F. This is adequate for most of the airplane, but they aren't used anywhere near the engine.

Instead, you'll typically see the AN363. By using metal fingers, the hole diameter narrows slightly at one end, and applies enough friction to prevent self-rotation.

Self-locking nuts aren't the only solution. If the bolt is intended to connect moving parts, the motion can overcome the nut's self-locking ability. Castle nuts

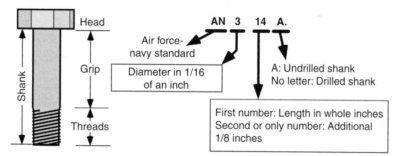

Fig. 12-5. *Specifications for AN-type bolts.*

are the answer. They look like conventional nuts with notches cut out of one side like a castle's ramparts. The notches pass a cotter pin through the hole in the nut. Castle nuts come in two flavors; AN310 for all application, and the low-profile AN320 for applications where the bolt isn't under tension.

Aviation cotter pins in various diameters and lengths are covered by the AN380 and/or the MS24665 specifications. One important point about them: *They cannot be reused*. If you take one off (allowed under Part 43), replace it with a new one.

So you'll want a collection of cotter pins handy whenever you work on your plane. Add that to some aviation washers and extra nuts of various types. You won't often have a need for spare bolts, but nuts have an annoying little tendency to bounce away and disappear. It's less irritating to have a batch of spares handy. Pick the dropped ones out of the dust pile next time you sweep the hangar.

Figure 12-6 shows what some of this hardware looks like and some part numbers for common sizes. For $60 or so, you can have a pretty good stock sitting around. You can keep them in 35-MM film containers, but cheap see-through plastic trays with compartments are better.

USING THE HARDWARE

There are a few basic rules to remember when bolting things together.

When you disassemble a component, make note of the orientation of the bolt head and the number of washers used. Most bolts will have their heads in the same direction; most use only one or two washers under the nut. If you don't note the configuration and orientation, here are some guidelines: First, bolts generally have their heads up or forward. This provides a "last ditch" defense if the nut somehow

	To fit AN3 (3/16 inch) bolt	To fit AN4 (1/4 inch) bolt	To fit AN6 (3/8 inch) bolt
Elastic stop nut	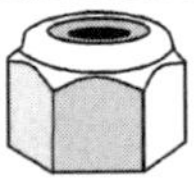 AN365-1032A AN364-1032A*	AN365-428A AN364-428A*	AN365-624A AN364-624A*
Metal stop nut	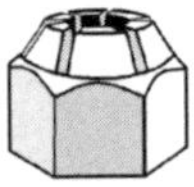 AN363-1032	AN363-428	AN363-624
Castle nut	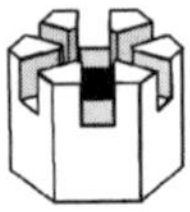 AN310-3 AN320-3*	AN310-4 AN320-4*	AN310-6 AN320-6*
Washer	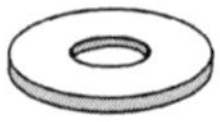 AN960-10 AN960-10L* AN970-3**	AN960-416 AN960-416L* AN970-4**	AN960-616 AN960-616L* AN970-6**
Cotter pin	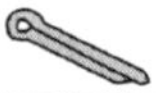 AN380-2-2	AN380-2-2	AN380-3-3

*Thin unit... use only when called for
**Extra-wide washer

Fig. 12-6. *Typical aviation hardware part numbers.*

comes off—gravity or positive acceleration will keep them in place. There are exceptions to this rule, usually brought about when a bolt installed in the normal way interferes with another component. Otherwise, take "head up or forward" as the standard. Take note of any bolts running the wrong way before removal. Restore them to their original orientation.

There's no similar rule for bolts running left and right, so memory or written notes will be necessary here as well. Copious notes, sketches, and digital photos will help, particularly on large projects or full-scale restorations where memories tend to fade. Besides, they're fun to look at later.

When tightening a bolt/nut assembly, always hold the bolt stationary and turn the nut. At least one washer should be used under the nut to keep it from grinding the surface of the component. When the assembly is tight, at least one thread of the bolt must be visible beyond the nut. If not, replace the bolt with a longer one.

If a bolt is *too* long, the nut may "bottom out" on the threads before it's tight. Try turning the nut without holding the head; if the assembly rotates, the bolt is too long. Remove the nut and add another washer. Don't use any more than three washers, though. If things still aren't tight, replace the bolt with a shorter one. *Do not* use a die to add threads to the bolt. AN bolts threads are rolled, not cut; cutting threads causes localized stresses that can cause premature failure. Also, cutting threads on an AN bolt removes the cadmium plating, giving corrosion a starting point.

These basic rules are summarized in Fig. 12-7.

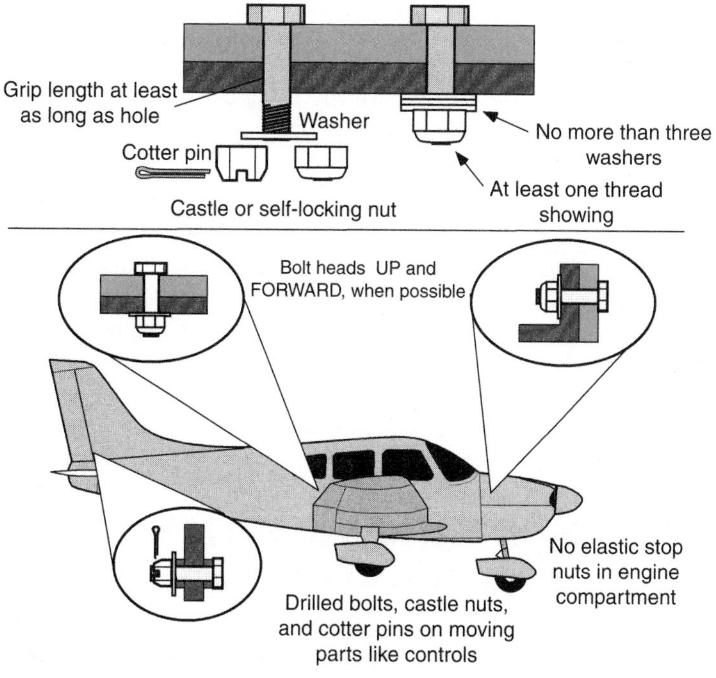

Fig. 12-7. *Basic bolt installation guidelines.*

Safety wire

Safety wire secures things that don't have self-locking features. Your oil tank's drain plug, for instance, will probably have to be safetied in place after an oil change. As will the oil filter or screen and a number of other small items. If it's safety wired when you start, you'll have to restore it when you're done.

Safety wire comes in several different varieties: soft iron, copper, brass, and stainless steel. Stainless is the most common. A variety of diameters are available: 0.020-, 0.032-, 0.041-, 0.051-, and 0.057-inch. Buy a spool of 0.032 inch and a spool of 0.041- inch wire.

Safety wire material is strong but not brittle—it can take a bit of flexing without breaking. Safety wire isn't designed to carry a load. It's used mostly to keep something from turning, usually, a nut. The safety wire will run from the nut to a solid point; sometimes a lug especially for the purpose. Other times it just runs to a convenient projection.

Before cutting off the old wire, make note of what both ends are attached to. You'll want to run the new wire to the same lugs. Discard the old safety wire; never reuse it.

The length of the new piece of wire will have to be more than twice the distance to be spanned, so cut off a generous hunk. If the piece is too short, it invariably doesn't become apparent until after you've spent 10 minutes stringing it through tiny little holes. Cut it long!

Safety wire is always installed so that it prevents the safetied component from turning in a loosening direction. This generally means that the wire should curve around in a clockwise direction, when viewed from the nut side. Bend the piece of wire in half, and slide one side of the wire through the hole of the item to be safetied. The ends should be pointing in the tightened direction. Next you'll have to twist the wire from the nut to the approximate distance to the safety point. This way, if one leg of the wire breaks, it's still wrapped with a solid piece.

The best way to twist the wire is with a set of safety-wire pliers. It's a slick little tool, albeit a bit expensive. The jaw ends grasp the two sides of the safety wire and the handle locks to hold them. Pulling a center knob spins the pliers and the attached wire. When correctly twisted, the safety wire should have between six and twelve turns per inch. Too loose, and the benefit is lost. Too many, and it may break from the strain.

When the first twist is done, release the pliers and run one piece of the wire through the lug. Then twist the wire beyond the lug. Use the nipper part of the pliers to cut off the excess wire, leaving at least five twists past the lug. Use the narrow end of the pliers to bend the end out of the way. This procedure is summarized in Fig. 12-8.

That should be enough basics to get you started. Now, on to the *real* work!

COMMON OWNER MAINTENANCE PROCEDURES

The following sections contain information to help you get started on the most common owner maintenance operations. However, I encourage you to find an experienced person to check your work the first couple of times. If your mechanic is

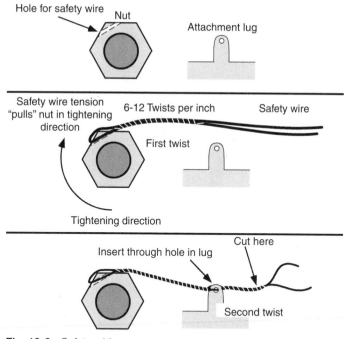

Fig. 12-8. *Safety wiring sequence.*

willing and convenient, great. Otherwise, check around for another owner doing a similar procedure.

A good source of unofficial assistance is your local chapter of the Experimental Aircraft Association (EAA). Homebuilders get quite experienced in maintaining their own planes, especially since the original builder of an aircraft can perform all the functions of an A&P.

One final caution: The simplified diagrams included aren't based on any particular design. They're just there to help you understand how the system works. Consult your service manuals for the diagrams appropriate to your own plane.

ENGINE OIL SYSTEM

From the owner's point of view, the oil system and the fuel system are two of the most important elements of the airplane. Why? Because a mistake during preventative maintenance can lead to sudden engine failure. So it behooves an owner to understand the whys, and not just the whats.

Oil system basics

A schematic representation of an aircraft oil system is shown as Fig. 12-9. This isn't representative of any particular brand of engine; rather, it's a collection of typical oil system elements that I'll use to explain system operation.

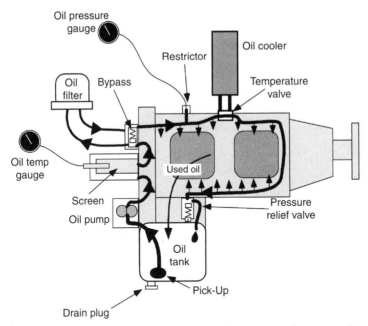

Fig. 12-9. *Simplified oil-system diagram. On some engine types, the pressure-relief valve is located in parallel with the oil pump.*

It all starts near the bottom of the oil tank. Oil enters the system through a pickup tube; this tube usually has a rather coarse strainer at the end to filter out large items that could plug the oil passages.

Suction to pick up the oil is provided by an engine-driven oil pump. The back of the engine contains gearing to turn a number of items, like the oil pump, magnetos, vacuum pump, and generator. It's called the "accessory case."

Once the oil leaves the oil pump, it travels through the oil screen. This is a rather coarse metal-mesh filter; Light Plane Maintenance magazine calls it the "elephant catcher." It gets the largish bits out of the oil, but none of the fine grit. The screen is a permanent item, not a throw-away element.

Obviously, engines stay healthier when the finer grit is filtered from the oil. Some engines include a spin-on oil filter similar to those on cars. These can attach directly to the engine, or be connected by hoses, a "remote mount" filter. Remote mount units are typically added-on after engine manufacture. The filters are much more effective than the screen. On most engines, the oil change interval *doubles* when a filter is installed. If you don't have time to change the oil yourself and must have an A&P do them, you can quickly make back the cost of the STC for a remote-mount filter.

One problem can occur, though. Since the filters have a finer mesh, they're more subject to clogging than the elephant catcher. To prevent a clog from stopping oil flow, aircraft filter attachments include a bypass valve, which switches the filter out of the oil flow if it becomes too restrictive.

A similar valve is placed at the base of the oil cooler, if one is installed. This valve works slightly differently; if the engine oil is cool, it shuts off oil flow to the cooler. As the oil warms, the temperature valve shunts more and more oil through the cooler. An A&P can adjust the valve to set the desired temperature.

Yet another kind of valve is installed either near the oil pump or well downstream. The pressure relief valve is used to control the overall oil pressure. Some systems block oil flow until it reaches the desired pressure, others shunt oil back to the input of the oil pump when the pressure attains the set point. If your oil pressure tends to run too high or too low, it might merely be a misadjusted pressure valve. Ask your A&P to check.

The FAA requires an oil pressure gauge in the cockpit. It gets its information from a tap in the oil flow. Lycomings generally put the *pickoff* near the oil pump; Continentals tend to put it well downstream. The difference can be seen on startup. A Lycoming will show oil pressure within seconds. A delay of up to 30 seconds isn't unusual in a Continental. It doesn't mean that the Lycoming has a better oil pump. If its pickoff were placed at the same relative location as the Continental, it would probably show the same behavior.

Oil pressure gauges are of two types. One places a transducer on the engine and sends the pressure to the cockpit gauge electrically. The second, more common way, is to use a mechanically-reading gauge and connect it to the engine oil system via a tube. If this tube were to break, however, oil would start pumping out of the engine. To minimize this hazard, a small-diameter tube is used or a restrictor is installed at the pickoff. The small size doesn't effect the pressure reading, but slows the rate of oil loss should the tube break.

It's important to understand the oil pressure system in your aircraft. If the tube breaks, the gauge will drop to zero. The engine still has oil; the break won't affect its actual oil pressure for a while.

A friend once lost oil pressure while flying over a large bay. At the same time, hot oil started spraying his legs. An A&P, he correctly assumed the gauge line had broken. He knew he had some time until the engine actually seized. He was able to reach an airport on a nearby island.

That's not to say that an indicated loss of oil pressure isn't an emergency. If the problem is a broken oil pressure gauge tube, the engine is still going to run out of oil, eventually. But if there aren't any other signs of distress (rise in oil temperature, strange noises, visible oil on the outside of the aircraft), consider pulling the throttle to idle rather than shutting the engine down. It might save the day should you start undershooting your forced-landing target.

Don't take chances with the souls on board, though.

Changing oil

With some small exceptions, changing aircraft oil is similar to changing oil on your car.

To begin, warm up the engine. Do a couple of touch-and-goes if the weather's nice, otherwise idle it for 10 minutes or so (away from other people's airplanes, thank you). Not only does warm oil flow better, but running the engine stirs up the crud so it'll flow out with the old oil.

Once you're back at the hangar, remove the cowling. Position a bunch of old newspapers under the engine and have a roll of paper towels standing by; you're likely to need them.

The oil tank contains a drain and drain plug at the bottom. You'll see two types of plugs. One is an automotive-style short bolt; just like a car, it must be unscrewed to drain the oil. The second is a short valve fitting that doesn't require tools. As shown in Fig. 12-10, it consists of a small metal assembly with a short metal tube jutting downward.

If you've got the plug-style, take your side cutters and remove the safety wire. Position a pan or pail underneath the tank. If you can rig a funnel with a short piece of hose, do so—oil tends to splash around. Take your wrench and loosen the plug. Once it's loose, remove it by hand. You should be able to sense when it's almost free. Make sure you hang onto it. Fishing for a dropped drain plug in a pan of hot, dirty oil isn't much fun.

If you've got the valve-style, rejoice. Put the pan under the engine, and slip a piece of garden hose long enough to reach the pan over the metal tube. When the valve is opened, the oil flows down the tube to the pan; no muss, no bother.

How do you open the valve? Generally, the collar around the fitting pops upward. It usually has to be turned 90 degrees, first. Ask your mechanic. Leave the oil to drain and drip while you take care of the screen and/or filter. Most engines have the screen

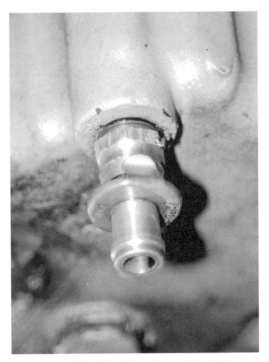

Fig. 12-10. *Oil quick-drain valves make oil changes easier and less messy.*

on the back of the accessory case. They can screw in or sit behind a bolted cover. Make sure you have a replacement gasket, if required.

Position a pail or some wadded-up paper towels under the screen location. Cut the safety wire and unscrew/unbolt the screen. You might need a pretty big wrench. Often the oil-temperature sensing bulb is screwed into the end of the screen housing. This has to be removed before the screen. On the small Continentals, hold the screen itself stationary with the big wrench while turning the bulb fitting. Once the screen is off, check the contents. Flush it with a little solvent (a tiny squirt of gas will do) and look for any metal pieces. Reinstall it and safety wire.

If your engine has a filter, you may not have to remove a screen. Check with your mechanic. While finding the screen, the first time your wanting to change oil might be a bit challenging, locating the filter is simplicity itself: Big, round, and (typically) white.

Before proceeding, take careful note on how the safety wire runs. Then clip it away and unscrew the filter. If it's reluctant to move, put a socket on the end hex fitting or use an automotive strap-type filter wrench.

Some engines don't like sitting around too long without an oil filter; they lose the prime in the oil system. So it's best to put on the new filter fairly quickly. First, clean off the filter mount on the engine. Make sure the old gasket is completely gone. Hold the new filter open-end-down and slap it a couple of times to knock out any junk that might have worked its way inside. Lubricate the end gasket with a little oil or Dow-Corning DC-4 lubricant (Fig. 12-11). If feasible, pour some new oil inside

Fig. 12-11. *Before installing the new oil filter, lubricate the end gasket.*

the filter. Careful, though—if the filter mounts on its side, much of it will just drain out again.

Position the filter in place and screw it on until the gasket contacts the mount (Fig. 12-12). The amount of torque required should be specified on the filter itself; most require about 18–20 foot-pounds. For those without a torque wrench or the right-sized socket, Light Plane Maintenance magazine recommends turning the filter for one and three-quarter turns after the gasket touches the mount. Safety wire the filter, and that job's done.

Now turn your attention to the old filter. It really should be cut open and examined. Cutters range from $60 to $150. Some owners just have the mechanic open the filter at annual time. There is an AD out on some engines that require cutting open the filter every 50 hours. If you've got one of those engines, the A&P will have to do the honors and make the appropriate notations in the logbook.

The second-to-last step in your oil change is to reinstall the sump's drain plug and safety wire it in place (or close the valve, if so equipped). Finally, add the right amount of oil to the tank. Not every aircraft engine wants to be full of oil. Many manufacturers recommend that the tank be filled to the full line only when the aircraft is about to leave on a cross-country. Engines include a "crankcase breather" line to vent any pressure in the crankcase. Some engines will happily spew oil through this tube as the airplane climbs or descends. Hence, the manuals call for a full oil tank only if the plane's going to stay at a constant altitude for while.

Sure, you can fill them to the "full" line, but you'll probably be scrubbing oil off the belly, later. Follow the instructions in your manual.

Another caution: Oil now comes in those little plastic jugs. Overall, it's a nice setup; we no longer need to chase down a metal spout and ram it through the top of the oil can. However, one little problem with those plastic jugs: When you open

Fig. 12-12. *Most aircraft oil filter spin-on just like car ones, except the filter must be safety wired in place. They should be torqued to 18–20 foot-pounds.*

them, a little plastic ring separates. It usually remains on the cap, but sometimes it sticks on the top of the jug. When it does that, it sometimes comes loose when the jug is tilted to pour oil into the engine. Sometimes it slides down the jug, into the oil filler, and into your oil tank.

Oh, oh.

Most oil jugs now have a split-ring, which stays with the cap. But if the ring stays with the jug, *take it off before pouring the oil.* Once the oil tank's full, replace the filler cap. Check that the plug is in place, and the plug, screen, and filter are safety wired. Then crank up the engine. Watch the oil pressure. It might be a little slower to come up, but not much. If it isn't off the peg by the manufacturer-stated time, shut down. Otherwise, run the engine for a few minutes, then shut down and look for oil leaks. If it's clean, button up the cowling and go for a short test flight.

FUEL SYSTEM

The fuel system is as important as the oil system. Even more so, perhaps. The FAA allows very little owner maintenance on it. You can "clean or replace the fuel . . . strainers and filter elements" and that's about it.

Like everything else, this is subject to interpretation. After all, most aircraft carburetors include a fuel screen deep inside, but I bet the FAA would prefer we didn't start disassembling our carbs.

Fuel system basics

A schematic of a low-winged aircraft's fuel system is shown in Fig. 12-13. Gas flows from the tanks, through the selector valve and gascolator, and into the carburetor, all the time hurried along by the main (engine-driven) and auxiliary (electric) fuel pumps.

High-wingers eliminate the fuel pumps and use gravity to move the fuel. The distance from the bottom of the tank to the carburetor is called the *fuel head*. There must be sufficient head to allow the system to provide 150 percent of the engine's full-throttle fuel burn. Otherwise, a fuel pump must be installed. Even high-winged aircraft may have a pump if they don't generate enough fuel head or a fuel-injected engine is installed. There are other permutations; some planes pump gas to a "header tank" forward of the cabin from where its gravity flows to the carburetor.

One oft-asked question: Why don't most low-winged aircraft's fuel selectors have a "both" setting? Two glasses, a couple of straws, and a bottle of your favorite libation will provide the answer. The glasses represent your tanks, the straws the fuel lines, and you get to be the low-winged airplane's fuel pump. Fill one glass, leave the other empty. Put a straw in each glass. Put your mouth over the other end of the straws, and suck on both at the same time. You get a mouthful of air. Pulling a liquid uphill is a lot tougher than sucking air. The pump sucks air instead of liquid. Use your tongue as a "selector valve" while considering how to simulate a high-winged plane. You'd take both glasses, the full and the empty one, and turn them upside down over your head.

Fuel will flow, all right.

There are, in fact, a few low-wing planes with "both" settings. But they use a bit more complicated system than just a single central valve.

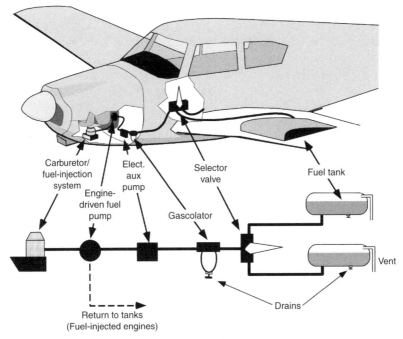

Fig. 12-13. *Simplified fuel system schematic.*

Fuel system maintenance

The main strainer and filter that you're allowed to work on is the gascolator, usually located low on the firewall. It's not something you'll have to mess with very often. For most owners, cleaning it at every annual will be sufficient. However, if you discover sediment or foreign matter in the fuel, open and clean the gascolator.

Begin the process by turning the fuel valve off. That "150% of full-power fuel burn" is no fun at all if it's cascading onto your hangar floor. Next, position a bucket under the gascolator to catch the gas that'll trickle out of the bowl and the lines. Open the drain on the gascolator to start the flow.

The gascolator bowl can be removed without taking the unit from the aircraft or disconnecting the fuel lines. There's usually a small nut or thumbwheel on the bottom on a metal bail (thick wire) slung underneath the bowl. Cut the safety wire on the thumbwheel and start turning. The bowl will gradually slide down. Sometimes it comes low enough to slide out without disconnecting the bail; otherwise, just pull the tops of the bail out from the upper housing. Often the gascolator doesn't have a bowl per se; top and bottom housings are separated by a 2-inch length of aluminum or glass tube.

The first time I opened the gascolator, I was sure I'd been gypped. There was nothing inside! I was expecting a big element like a car fuel filter. A careful examination found a skinny stainless-steel screen underneath the housing (Fig. 12-14). Other types do have a larger canisterlike filter screen.

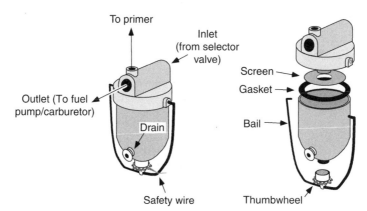

Fig. 12-14. *Typical gascolator. Many aircraft use a canister-type filter instead of the small stainless-steel screen.*

Remove and clean whichever type you have. Wipe out the inside of the bowl and the top housing. Reassemble the unit the same way it came apart, making sure the neoprene or cork gasket atop the bowl is properly positioned inside the upper housing. Tighten the thumbwheel and safety-wire the assembly. Turn on the fuel and check for leaks. Run the engine for a minute or two and check again. A leaky gascolator is dangerous. Not only can the loose fuel catch fire, but air can be introduced into the fuel line to the carburetor. Engines don't run well on air.

In addition to the carburetor screen mentioned earlier, there are other screens in the fuel system. Each tank, for instance, includes a "finger strainer" at the outlet port. You probably won't get involved with these unless you're restoring the aircraft.

As a side note, the mesh on the fuel system filters gets progressively finer from the tank to the carb. The finger strainer is relatively coarse, the gascolator is finer, and the final screen within the carb is fine indeed.

WHEELS AND BRAKES

As mentioned earlier, owner maintenance on the wheels and brakes is another interesting Part 43 interpretation issue. The FAR doesn't include any brake servicing under preventative maintenance. Yet, if your plane has disc brakes (like most aircraft), you'll have to remove the brakes when changing a tire. But when the brake assemblies are off the landing gear, it's often a dead-simple job to replace the brake shoes. Yet the regulation doesn't allow you to.

Why use such a system, anyway? You don't have to remove the brakes on a car if the tire goes flat!

It all boils down to saving weight. A disc brake works by clamping two pads on either side of a metal disc attached to the spinning wheel. The disc on a car is attached to a hub, which bolts to the axle and incorporates the necessary bearings. The wheel then bolts to this hub. A flat tire won't strand you outside East Podunk; you can quickly unbolt the wheel and slap on a spare.

Rather than a separate hub and wheel, aircraft save weight by combining them in one light unit. It's not a major concern, safety-wise. A flat tire on your plane doesn't leave you stranded in a cow-skull-studded desert. In most cases it happens at a nice, safe, airport.

Disk brakes are pretty much standard now, but some older planes have drum-type brakes. Wheels on these can often be removed without disturbing the brake system.

You're allowed to replenish brake fluid. Sometimes the reservoirs are easily located (Fig. 12-15) and sometimes they're buried under the instrument panel as part of the brake master cylinders themselves. One no-mess system for adding brake fluid is by using a pump-type oil can. One warning: Use only aviation brake fluid. The automotive stuff is *not* compatible with aircraft systems.

Before we get started, let's get our nomenclature straight. When I say "tire," I mean the stiff black rubbery thing that's in contact with the asphalt. *Wheel*, technically, is the metallic thingamabob the tire is wrapped around. I'll use the term *rim* for the edge of the wheel closest to the outside, where rubber meets metal. The *tube* is the flexible rubber inner tube. The *valve* is the part where you add air; the *stem* is the extrusion from the tube that holds the valve.

Now, let's go through wheel removal and disassembly.

Jacking

The process begins with figuring out how to jack up the airplane.

Fig. 12-15. *The brake fluid reservoir of this Piper Arrow is easy to get to; not true of many aircraft.*

Park the airplane on a flat spot away from grease spots or loose gravel. The wheels that aren't going to be jacked should be chocked fore and aft. Next, remove any safety wire or cotter pins, ensuring you've got replacements handy. Sometimes it takes a lot of jerking and pulling to get the cotter pins out. For safety's sake, do it before the airplane is up on jacks.

A variety of formal and informal methods are used to jack the airplane. Your service manual will show the approved methods (Fig. 12-16). Usually, there'll be "hard points" where a standard hydraulic bottle or floor jack can be placed. Sometimes the tiedown ring is identified as the hard point; it's unscrewed and replaced with a special jack fitting (Fig. 12-17).

Other planes use special adapters. Bonanzas were delivered with fittings that slid into the back side of the gear leg to allow jacking. If a tall wing jack isn't available for your Cessna, various adapters clamp to the gear legs. If you're changing a nosewheel, a tight tail tiedown rope might be just the ticket.

Before you hoist away, though, consider: What's going to happen if the jack slips, or suddenly goes flat? When the wheel comes off, slip a block of wood, stack of 2 × 4s, or old milk crate under the bare axle, just in case (Fig. 12-18).

Hoist away when ready. Watch the base of the jack; make sure it doesn't start sliding. If you're jacking spring-steel (Cessna) or bungee-style gear legs (early Piper, Aeronca), be wary. The leg will tend to splay outward a bit and tilt the jack. Use a rubber mallet or piece of wood to tap the base of the jack back into place.

One last point: Be careful when jacking taildraggers. If you place the jack behind the wheel, the plane can tip forward on its nose.

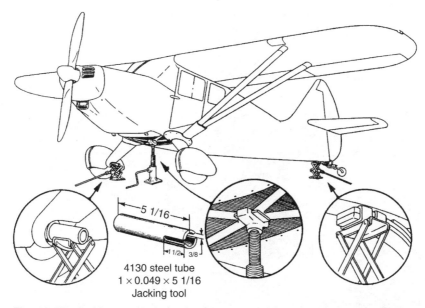

Fig. 12-16. *Jacking methods for the Stinson 108. (Courtesy Univair Aircraft Corporation.)*

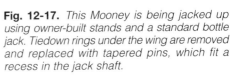

Fig. 12-17. *This Mooney is being jacked up using owner-built stands and a standard bottle jack. Tiedown rings under the wing are removed and replaced with tapered pins, which fit a recess in the jack shaft.*

Fig. 12-18. *Whenever the wheels are off, place something under the axle to prevent damage to the plane should the jacks slip.*

Fig. 12-19. *When the through-bolts are removed, the caliper can be lifted away from the wheel. Note the light-colored brake pads.*

Removing the wheel

Wheel removal is fairly straightforward. The primary thing is to remember the order it came apart. Draw a sketch, shoot pictures, or lay out the parts in sequence.

The hardest part is getting the brake calipers out of the way. The caliper is a U-shaped device wrapped around the brake disc. It's held in place by (typically) two bolts. Some planes have rubber brake lines; on these the whole caliper can be unbolted and removed (Fig. 12-19). It shouldn't be necessary to remove the brake hose, but use a bit of tape or wire to hold the caliper so that it's not just hanging on the hose.

With planes having steel lines, you'll split the caliper in place. Whichever way, there may be some metal shims between the two halves of the calipers—remember which goes where.

Wheels are held in place by a variety of methods. Those installed on forked legs usually have a long bolt through bushings in the fork. Otherwise, the end of the axle is probably threaded for a large nut. These are shown in Fig. 12-20.

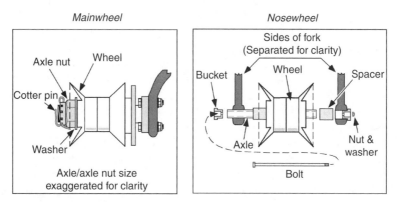

Fig. 12-20. *Typical mainwheel (left) and nosewheel (right).*

When ready, remove the axle bolt/nuts and other parts like washers, taking careful note of their order. At that point, the wheel should be free. Slide and wiggle it off the axle.

Removing the tire

Automotive wheels are one-piece affairs. While simple and cheap, they complicate the removal of the tire from the wheel. If you've watched a tire shop at work, you've seen them use a *bead breaker*. They ram this angled piece of steel between the inner part of the tire (the *bead*) and the wheel, and use a motor-driven system to drag it around the wheel. It pops the bead to the outside of the wheel.

Aviation? We just take the wheel apart.

Start by marking the two sides of the wheel so they'll go together at the same orientation. This'll make sure it's still in balance when reassembled. Mark the brake disc, as well. The tire will have a stripe or dot painted on it to indicate the light side; it is to be aligned with the valve stem when the wheel is reassembled.

Next, let the air out of the tire. The easiest way to do this is to remove the valve core. The simple little tool to do this is available at any auto, bicycle, or hardware store. Pick up a pack of replacement cores, too. Often a slow leak is caused by a faulty core; you can replace the core and pump up the tire without removing it from the airplane.

Once the air's out of the tire, you'll need to separate the bead away from the wheel's rim. All you have to do is get the tire loose—you don't have to flip the edge of the tire over the rim.

To break the beads on my Fly Baby, I use four simple, commonly available tools: two knees and two palms. I set the wheel flat on the floor and kneel on the tire, evenly spacing my knees and palms around the circumference. Then I kind of jiggle up and down until the tire pops away from the rim. Flip the tire over to separate the other side. However, I'm a tad heavier than many of you. If worst comes to worst, try standing on the tire with your feet on either side. Carefully hop until the bead breaks away.

You can avoid this hassle with a bead breaker. The aviation variety is a lot more civilized and cheaper than you'll find at the local gas station. What you *don't* want to do is stick anything sharp (like a screwdriver) between the rim and the tire. You'll probably gouge the wheel, and quite possible puncture the tube. Tubes can be patched but wheel gouges can lead to cracking.

From this point on, it's easy. Most aviation wheels consists of two halves bolted together (Fig. 12-21). While similar, one half will have a hole for the tube stem. Note whether the bolt heads or the nuts go on the stem side, then unbolt. The tire will slide clear of the wheel halves. Be careful not to snag the stem. If you've got disc-type brakes, the disc comes off with the bolts.

Other types of wheels come apart similarly. Cub wheels, for instance, have a snap ring on one side; once a retaining pin and the snap ring is removed, the end rim slides right off. When the tire's off the wheel, reach your hand into the opening and carefully slide the tube out. Do this slowly; even with the valve core out, the tube is sometimes reluctant to collapse down.

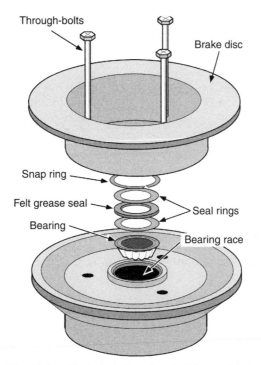

Fig. 12-21. *Exploded view of typical General Aviation wheel. Larger wheels use six through-bolts.*

Wheel bearings

Anytime a wheel's disassembled, you might as well inspect and regrease the wheel bearings. Generally, the service manual will call for regreasing at every annual. To some extent, it depends on how much your plane taxis.

One part of bearing work appeals to the kid in all of us: You get to play with lots of messy axle grease. However, keep the grease away from the tire and tube. Petroleum products hasten rubber deterioration.

Generally, a snap ring holds the bearing and spacers in place. Remove the snap ring with a needlenose pliers or the tip of a screwdriver. Take off the spacers and any felt grease retainers, taking care to remember the order. I just stack them in sequence and reinstall all at one time.

Cleveland wheels use a slightly different system. The rings are held on with three small screws. Removing them frees the assembly. The bearing should slide out of its race. Wipe off the surface grease and take a close look. Your inspection shouldn't reveal any flat or loose rollers or any corrosion. Any cracks or deformations mean that it's time to get a new bearing. Ditto if the bearing grates when turned or the rollers seem loose. If you need a new bearing, try to read the make and part number off the old one. Call a local bearing distributor rather than an aviation parts shop.

Now comes the fun, greasy part: repacking the bearing. Some references call for cleaning out the old grease with solvent. Not a bad idea, but one that's fraught with danger. You must ensure the bearing is *completely* dry before packing new grease into it. Any drops of remaining solvent will keep the new lube from sticking—not good for the bearing. I generally just clean off the surface and pack new grease in until the old stuff is forced out.

Open your can of grease (check your service manual, but it'll probably be the same stuff sold at the auto-parts store) and slap a large dollop into the palm of one hand. Cup your palm around the mound. Hold the bearing in your other hand, and press the tapered (roller) side into the grease (Fig. 12-22). The new grease is forced in between the rollers. You should see a little bit of darker old stuff ooze from between the inner and outer race. Repeat until new grease oozes out. Then rotate the bearing a little bit to bring another set of rollers around, and shove it into the grease again. Work the bearing around, adding more grease to your palm when necessary.

Fun work. I'll be honest, though—there are tools that'll let you pack the bearings with an ordinary grease gun.

When done, clean the grease out of the race and slip the bearing into position. Wipe your hands clean on a paper towel, then restore the seals and spacers to their position. Flip the snap ring into place, and then repeat the process on the bearing in the other half of the wheel.

Wheel reassembly

Before putting things back together, it's a good time for an inspection. Wipe everything clean. There shouldn't be any grease visible; the bearings depend on the grease inside, not outside.

If you disassembled the wheel due to a flat tire, the problem undoubtedly lies in the tube. Happily, there's no magic technology involved, any old tube repair kit

Fig. 12-22. *Greasing wheel bearings is easy, but messy.*

will suffice. Have a new tube standing by, in case the old one is unfixable. Start the reassembly by slipping the tube back into the tire. It's a tight fit, but don't spray lube on it! Many so-called "silicon" sprays include oil, too.

Instead, use baby powder. Dash it liberally over the tube and inside the tire. Bunch up the tube and slip it into the tire. Reach in from the other side to pull (gently!) and guide it into position. If it starts to jam, dump some more baby powder on it.

Check the tire indicator (white stripe or red dot) and slide the tube around until the indicator is properly aligned with the tube's inflation valve. On my tires, the dot must be aligned with the valve. Some tubes also have stripes painted on them; use this stripe as a reference instead of the valve.

Once the tube's in place, pump it up just a little bit, then let the air out again. This gets the last little creases and whatnot out of the rubber. Slide the wheel half with the valve cutout into the tire. Place that side down, then slide the other half into place. Look carefully at the mating surfaces of the wheel halves. They should be meeting exactly, if there's a gap, it could be because the tube's pinched. You should be able to hear the metal wheel halves grind against each other.

Rotate the stemless wheel half until your alignment marks coincide, then bolt the wheel together. Don't forget the brake disc. Once the bolts are all lightly installed, tighten them gradually with a cross pattern just like the lug nuts on a car. The wheel should have the appropriate torque values stamped on it.

Once everything's tight, inflate the tire to rated pressure, then let the air out. This, again, allows the tube to shift around and get comfortable, which is why we put baby powder inside the tire as well as on the tube. Then install the valve core and reinflate.

Reinstalling the wheel

With the wheel reassembled and inflated, it's time to put it back on the airplane. Wipe a little light oil over the axle to lubricate and slow corrosion. Slide the wheel into place and reassemble the axle assembly. If the manual shows an appropriate torque for the axle nut, go ahead and apply it. Otherwise, tighten the nut while spinning the wheel. You want it tight enough so that there's just a little drag. The wheel should coast a little when you let it go, but should drag down to a halt rather than spin on forever. Most people tend to install the wheel too loose, not too tight.

Install a cotter pin (rotating the nut backward slightly, if necessary, to get the castle nut lined up with the hole in the axle) and reinstall the brake calipers.

The next time you come out to the airport, have your compressor handy. Newly-assembled wheels will tend to lose a little pressure as the tube stretches into place.

Rotating the tires

Since they're used only for a few minutes per flight, plane tires don't wear out as fast as car ones. My plane goes three or four years before new tires are necessary. But it takes a little action to get them to last that long. On many aircraft, unless the suspension is perfectly set up and matched to your typical flight load, tires will develop unusual wear.

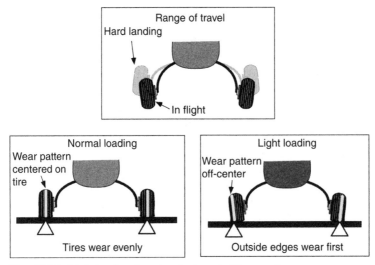

Fig. 12-23. *Light loading (or misadjustment of the gear) can cause uneven wear of main gear tires.*

Figure 12-23 shows a typical Cessna spring-steel landing gear. To absorb the shocks of landing, the gear legs are designed to move through a considerable arc. The legs hang down somewhat in flight, but during a hard landing, they'll splay out considerably.

Tires last the longest when they run perfectly vertical to the ground. But this condition can only be achieved at a particular load level, usually at gross. If you fly your 172 solo most of the time, your landing gear will tend be a little bowlegged. The tops of the main gear tires will be farther apart than the bottoms, a condition tire people call "positive camber." It's harmless, from a control point of view. Unfortunately, as Fig. 12-23 illustrates, camber moves the wear areas of the tires outward. This results in once side of the tire wearing more than its other side.

The solution is obvious—periodically, you've got to rotate the tires to let the other side wear for a while. Just changing wheels from left to right won't do it. The tire must be physically removed and flipped over, as Fig. 12-24 shows. Remove and separate the wheels, and reassemble with the opposite side of the tire on the stem side.

About that brake disk

Most disk-brake equipped airplanes, like cars, come with steel brake disks. However, unlike cars, airplanes don't use their brakes that often. The steel disks will tend to rust between uses.

If your plane flies regularly, or the local climate is dry, this won't be much of a problem. Otherwise, though, the surface of the disks will become roughened with corrosion. This isn't so much of a safety problem as a brake-wear problem. The rough surface will tend to eat your brake disks. They'll require replacement much more often. In the two years I owned my 150 in the damp Pacific Northwest, the brakes were relined twice.

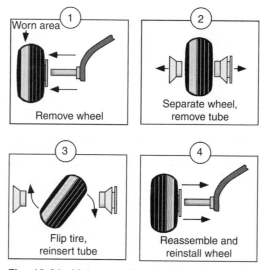

Fig. 12-24. *Make your tires last long by rotating them occasionally.*

If you have a similar problem, consider replacing your steel disk with a set of chrome or stainless-steel ones. Your mechanic can give you more information.

IGNITION SYSTEM

There's one allowed owner-maintenance function on the ignition system: removal, cleaning, gapping, and replacement of spark plugs. Still, you should know a bit about the overall system.

The magneto

I have a confession to make.

Several of my friends installed automotive engines in their homebuilt aircraft. One of their reasons was to eliminate that "hard-starting, trouble-prone, 70-year-old-technology magneto." Their engines all include modern all-electronic solid-state ignition systems.

My secret? I *like* magnetos!

Yes, they're old technology that Glenn Curtis probably laughed at.

Yes, electronic ignition allows tailoring the spark to the running condition; retarding spark to help starting or advancing it for maximum power. But when you hold a magneto, your hand is cupping your plane's *entire* ignition system. It doesn't depend on a battery. It can't be stopped by a burned-out solenoid or a shorted-out power bus. It's not a plastic box crammed with static-sensitive integrated circuits and connectors with corroding pins. A magneto contains a magnet, a rotor, a coil, a breaker switch (points), a capacitor, and that's it (Fig. 12-25).

Fig. 12-25. *While old technology, the aircraft magneto is a compact all-in-one system.*

The engineer in me *loves* such simple solutions. And a magneto is as simple as an ignition system can get. Sure, they aren't perfect, that's why the FAA makes us carry two of them. But dual failures are vanishingly rare. If the engine quits dead, suddenly, the failure is probably somewhere other than the ignition system. Of course, if the engine just won't start, there's a good chance the problem is with those dad-blasted magnetos.

Magnetos operate very similarly to a traditional automotive ignition system, with one important difference: A magneto generates all ignition power internally, rather than needing juice from a battery or generator. Figure 12-26 shows the operation of a basic magneto. The unit bolts onto the accessory case and engages a gear located inside. The internal shaft of the magneto turns three things—the magnets, the cam, and the rotor.

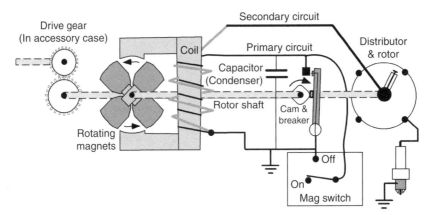

Fig. 12-26. *Basic magneto schematic.*

The coil is wrapped partially around the rotating magnets. Their rotation induces current flow in the coil, just like any generator. This current flows in the primary circuit, through the closed breaker points and back to ground. This current through the coil produces a magnetic field. The secondary circuit is affected by the primary's field and builds its own voltage in response. However, its voltage is dependent upon the rate of change of the magnetism of the primary, which doesn't really change very fast.

But rotation of the shaft turns a cam that opens the breaker points and disrupts the flow of current in the primary. Its magnetic field collapses very quickly; the secondary picks up the change and amplifies it. This high voltage is transmitted to the appropriate plug by the rotor and distributor.

When the breaker points open, the voltage in the primary would normally just arc across the points. That's where the capacitor (or condenser, to a traditionalist) comes in. It acts as a sort of high-voltage temporary battery and momentarily absorbs the voltage. By the time it's fully charged, the points are too wide to arc across. Without the capacitor, spark power is lessened and arcs tend to scorch the breaker points. Without the capacitor, the strength of the spark is severely lessened.

Just like a car, timing is critical. The magneto produces sparks at intervals; if the cylinder isn't at the appropriate position at the time the plug fires, efficiency is lost. The magneto body itself is rotated to adjust this timing. You'll note a slotted track for the mag's hold-down bolts. This allows your mechanic to loosen the mag, adjust it to the right timing, and tighten it down in that position. Unlike a car, this is not done with the engine running, and a conventional "timing light" isn't used. Timing the magneto is not an owner-maintenance procedure. Bad timing can cause immediate, severe internal damage. Leave it to your A&P.

As a pilot/owner, your main interface with the magneto is turning it on and off. Take a look at the figure again; see how the mag switch is wired. When you flip it to the "OFF" position, you're actually grounding out the breaker points. The magneto can't fire, because (electrically) the points never open. That's if the mag switch is operating properly. If the switch or its wire breaks, the pilot cannot ground out the breaker. The mag is always "ON;" a plug could fire if the prop was turned.

From the pilot's point of view, this is great. An in-flight failure of the wire or switch won't kill the mag. From the owner's point of view, it stinks. It means the engine could "kick" anytime the prop is turned. *Always* treat the prop like a mag is hot. From the owner-who-loses-his-keys-at-a-remote-airstrip's point of view, things are somewhat rosier. Disconnect the single wire running to each mag from the mag switch, and handprop the engine to start.

One drawback of magnetos is directly related to the fact that they generate their own power. Since the mags are turned by the engine, they produce a weaker spark the slower the engine turns. And the starter turns the engine at only 200 rpm or so.

To help, some mags have a booster to generate a hotter spark at slow RPMs. "Shower of Sparks" is the trade name of one such scheme. Retarding the spark (making it later) helps starting at slower speeds. Most airplane engines have at least one mag that retards the spark at low RPMs. You'll hear a click or "snap" as the engine's turned over slowly; that's the sound made by the impulse coupling.

Like the old-fashioned automotive ignitions they resemble, magnetos are subject to a number of ills. Breaker points burn, condensers open or short. Coil problems are tricky. Often the coil will work well when cold (or in denser atmosphere) but develop an open circuit when they warm up, or get to altitude. A mag that checks out well during a runup at home may show a miss after a short layover at an intermediate field, and then work perfectly when the mechanic finally arrives to look at it. When you suspect magneto problems, take careful note of the exact conditions—temperature, altitude, time of flight, and so on.

Spark plugs

The one bit of ignition maintenance an owner is allowed to do on an aircraft is remove, clean, regap, and install spark plugs. In addition to the deep socket mentioned in the tool list, you'll need a set of replacement spark-plug gaskets. These cost a quarter each, less if you buy a whole box.

Before removing your plugs, blow the loose dirt and scale away. If you don't have an air compressor, at least puff through a piece of rubber tubing held around the base of the plugs. You don't want any loose material to fall into the open hole.

While the general theory of the aircraft ignition system is similar to that of a car, components are a bit different. Everything's a bit more rugged. The ignition harness for most airplanes is heavily shielded to prevent radio interference. It screws onto the plugs rather than just clipping to them (Fig. 12-27).

To remove a plug, first loosen the harness nut atop the plug. When loose, pull it straight out. The harness terminates in a ceramic cylinder, called the "cigarette," with a small spring at the end. Look, but don't touch. Finger oils can cause contact

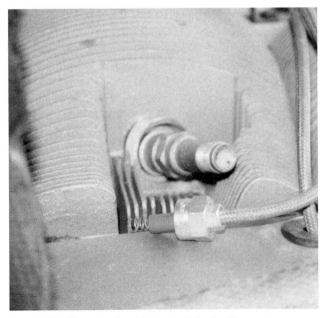

Fig. 12-27. *Plug wires for shielded aviation ignition systems screw onto the top of the spark plug.*

problems and misfiring. Examine the spring carefully; if discolored by corrosion, clean it off with sandpaper.

To take the plug out of the engine, you'll need your deep 7/8-inch socket. Clip it to your wrench, and slide the socket down the plug until it's well seated.

Plugs are installed to a particular torque. So they'll need a bit of a heave to get them to start turning. Don't just wrap both hands around the end of the wrench and pull, though. The socket is so long that just pulling on the handle can apply some bending load to the plug, as well. Instead, support the head of the wrench with one hand while applying torque with the other, as shown in Fig. 12-28.

When the plug loosens, use the ratchet to back it the rest of the way out. If you're going to pull all the plugs at this session, have a muffin pan or egg carton to set them in. Keep each plug separate and label them to indicate, which cylinder they came out of, and whether it was the top or bottom plug. If you drop one, throw it away. Unseen damage can rear its head later.

While the layout of the electrodes at the end of the plugs is a bit different from your car, they're the same basic design. The ceramic insulator shouldn't be cracked or have any pieces missing. The electrodes should be intact, not bent or melted. A brownish-gray deposit around the electrodes is the sign of normalcy. The center electrode starts out circular, but goes gradually out of shape with wear. Black deposits indicate carbon or lead fouling. If the deposits look oily, it could have a ring or valve problem in that cylinder.

If any plug seems abnormal, it's time to talk to your mechanic. Remember which cylinder(s) the bad ones came from.

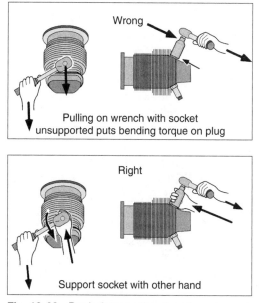

Fig. 12-28. *Don't damage the plugs by using only one hand on the wrench.*

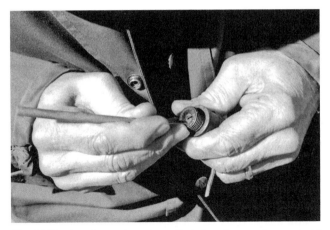

Fig. 12-29. *Lead or carbon deposits in the plug can be removed with a fine tool like a dental pick.*

One all-too-common problem you can alleviate is lead fouling. Many engines designed to run on 80-octane fuel have trouble with the lead contained in 100LL. This lead tends to clump in the plug, forming little "nuggets." Carefully scrape them away with a fine tool, like a dental pick (a screwdriver is too clumsy), as Fig. 12-29 shows. If you have access to a spark plug cleaner, a little easy blasting should help.

The lead causes more problems than crudded-up spark plugs, of course. Your only solutions are to buy an auto fuel STC or use Alcor's TCP in your fuel. TCP doesn't do anything magical to the lead, it just ensures that the lead deposits are softer. It helps the engine expel the lead rather than form solid clumps on the inside of the combustion chamber.

Use a soft brush to eliminate any loose particles on the plug prior to setting the gap.

I used to adjust the gap on my '74 Volkswagen by pressing the electrodes down against the fender. Airplanes require a bit more finesse. Pick up an aviation plug gapping tool. Prices vary, but a good-quality one can be found for $20 or less. You'll also need a gap feeler gauge, for about the same price. Practice a bit on an old plug (or get some dual from an old mechanic) prior to risking your flight plugs.

Before reinstalling the plugs, you may want to shift them around to equalize wear. Bottom plugs should become top plugs and vice versa. Shift them diagonally between cylinders, too—the top plug on the front left cylinder should be moved to the bottom hole in the back right one. Your engine manual should show the proper rotation. The reason for shifting from the top to the bottom is pretty obvious. But why switch between cylinders?

The magneto is essentially an Alternating-Current (AC) generator. Relative to the aircraft ground, it produces a high voltage positive pulse, followed by a negative one. The way the distributor works, each cylinder gets the same polarity each time. Switching the plugs around changes their firing polarities and evens the wear.

There's one very important last step prior to reinstalling your plugs. They're subjected to vicious extremes of heat and pressure, which can tend to lock them in place. This puts you in a world of hurt the next time you want to inspect the plugs.

To prevent this problem, coat the second and third plug threads with a special lubricant called *anti-seize*. It's available through aviation parts sources. Careful not to get any on the electrodes, and don't use it on the nut that attaches the harness.

Slide a new gasket onto the plug and start screwing it into the cylinder by hand. Put your socket on a torque wrench and tighten the plug to the torque specified in your service manual (usually around 28–30 ftlbs).

Clean the cigarette with a little solvent (acetone or unleaded gas), let dry, and slip it into the plug. Tighten up the nut, and you're done.

ELECTRICAL SYSTEM

For those familiar with car electrical systems, this should be old home week. Those with fumble fingers should also rejoice; most of the allowed preventative maintenance involves little more than changing a bulb or two.

A simplified diagram of a typical aircraft electrical system is shown in Fig. 12-30. It all starts with the battery. There are several different types. Some are practically indistinguishable from the lead-acid units used in cars. Other batteries feature a gel-type electrolyte, which replaces the liquid acid with a substance that doesn't vent gas when charging. These batteries are completely sealed and require little maintenance.

The negative terminal is attached to the aircraft's metallic structure, *grounded* is the technical term. Most devices on the aircraft can then connect to the negative terminal by simply achieving a good connection to the metal structure.

The positive terminal of the battery goes to one contact of the master solenoid. Turning the master switch "On" doesn't physically connect the battery to the aircraft power system. That would require a heavy-duty switch in your panel with attendant long runs of heavy battery cable.

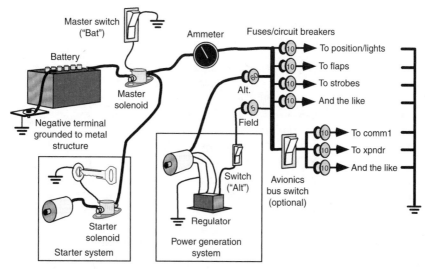

Fig. 12-30. *Simplified electrical system schematic.*

Instead, the small master switch in your panel activates the master solenoid, which includes the thick copper plates necessary to work with such current. Other terms for solenoids include "contactor" and "relay." It's the source of the big "Clunk!" you hear when you turn on the master.

The master solenoid brings enormous safety advantages. My dad's car burned up in an electrical fire. If he'd been able to disconnect the battery early, the fire probably wouldn't have happened. That's what the master solenoid is all about. Turning off the master switch disconnects the battery from most of the airplane. Exceptions include the clock, which is protected by its own fuse.

Two heavy cables emerge from the outlet side of the master solenoid. One goes right to another solenoid, this one for the starter. On most airplanes, the key activates the starter solenoid, which then applies power to the starter motor itself. The other cable goes to the main electrical power bus and to the aircraft's generator or alternator system. The power for individual electrical-driven components is run through a fuse or circuit breaker for protection. On some airplanes, a separate switch is provided for an avionics power bus. This protects sensitive electronics from the power variations produced during starting or stopping the engine.

To provide power while the engine is running, the aircraft includes either a generator or an alternator. The difference isn't really important, but most modern planes have an alternator since they're better at producing power at low RPMs, don't have heavy current brushes and commutators to produce radio noise, and are much lighter.

Electrical woes

Aircraft electrical systems seem to be involved in more than their share of problems. Let's take a look at some of the causes, and what you can do to mitigate them.

Let's start with the battery. As mentioned earlier, the negative terminal is attached to the aircraft structure. This avoids having to run two wires to all the electrical devices; they can obtain contact to the negative terminal by hooking onto the metal structure.

Simple and lightweight, grounding in this fashion can lead to difficulties. Engines are usually mounted on insulating rubber shock mounts, hence you'll see a *ground strap* running to the engine block to ensure its grounding. Other areas may feature such straps as well; the bare braided metal they're made of is easily recognizable.

Loose ground straps can cause irritating electrical problems. Things just don't work; they flicker and develop strange anomalies. If weird electrical problems arise, start checking the ground straps. Check that they're solidly bolted in place, and that the braid is solidly connected to the terminal. You might even unbolt them and clean the contact surfaces with a file or emery cloth. This isn't included on the list of preventative maintenance, but as long as it's a simple connection, the FAA shouldn't mind.

Another source of recurring problems is the master and starter solenoids. When you turn the switch on, you're applying power to an electromagnet. As shown in Fig. 12-31, this pulls in a soft iron core attached to a copper bus bar, which

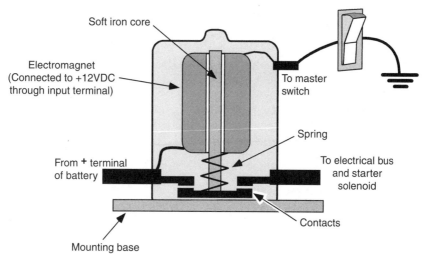

Fig. 12-31. *Workings of the aircraft solenoid or contactor.*

connects the input and output of the solenoid. A break in the rather fine wire of the electromagnet's coil will keep the solenoid from activating. And while the solenoids are usually sealed, corrosion still seems to appear. It can prevent the contacts from making good contact, or can jam the core so it can't move.

Unfortunately, solenoids can also fail in the inverse manner; they can activate and refuse to turn off. If this happens on the starter solenoid, you can be in deep trouble and not even know it. The engine's running, but the solenoid keeps the starter gear engaged and tries to keep turning the starter. If you don't notice, it can quickly chew up the starter Bendix and starter ring gear. On some engines, shavings from the Bendix and ring gear will fall into the oil flow, which then spreads this abrasive poison through your engine.

Deep trouble, like I said.

You may be able to hear it; a sort of rattle from up front. Depending upon your particular electrical system, the ammeter could provide another clue. Look back at Fig. 12-30. The ammeter isn't positioned to tell the pilot what the starter's current draw is. But if the starter keeps turning, the generation system will try to supply power for it, this will show up on the ammeter as a high rate of charge. The alternator thinks it's charging the battery, when in reality it's powering the starter.

Indications might be different on your engine, though. Talk to your mechanic. The important thing is to shut down the engine and switch off the master if you suspect the starter is still powered.

Problems with the master solenoid are minor in comparison, but still irritating. If the master solenoid stays on, it'll drain the battery. If the battery's fully charged but the solenoid won't activate, you still don't have power.

There's even the potential for combining these problems. Corrosion can not only keep a solenoid from working; it can provide a high-resistance path for slow current flow even when the master switch is off. If you fly the plane each day, you don't notice anything wrong. But if you only get out to the airport every two weeks or so, the battery has enough time to discharge through the high-resistance path. The battery's dead.

Often, your immediate reaction is to replace the battery. Over a hundred bucks' worth of new battery later, the problem is still there. I know. I went this route.

Before making the same mistake I did, check out the solenoid. If you've got an ohmmeter, disconnect the battery and measure the resistance across the solenoid terminals. It should be infinite. If you haven't got an ohmmeter, just charge up the battery and disconnect it from the aircraft before you leave. If the battery still has enough power the next time you come out to fly, the problem probably lies in your master solenoid. If it's flat, it's time for a new battery.

Battery maintenance

Maintaining your aircraft battery is fairly simple.

After removing the battery-case cover (if any), you'll be facing the top of the battery (Fig. 12-32). The terminals are different from those on cars. The aircraft cables actually bolt to the battery rather than clamping around the terminals.

Fig. 12-32. *Most aircraft batteries are contained in boxes under the cowling. Note the lift strap for removal.*

If you've got a conventional lead-acid battery, unscrew the caps and check the electrolyte level. The liquid should be at the base of the rings. If the level is low, add distilled water to bring it up to the mark. Replace the caps. Take a wad of paper towels, sprinkle baking soda on them, and wipe the liquid away from the top of the battery. Yeah, it's probably just spilled water. Use the baking soda anyway. Make sure the battery caps are already on.

Corrosion can work its ugly way between the terminals and the power cables of any type of battery. You should disconnect the cables on occasion and service the terminals.

Take a look at the terminals. The positive should be marked with a plus (+) and the negative one with a minus (–). Disconnect the *negative* terminal first. Remember, it's grounded to the aircraft; every piece of metal on the plane is electrically the equivalent of the minus side of the battery. If you tried to disconnect the positive terminal first and your wrench slipped and touched metal, you'd throw one heck of a spark. Not good for the battery, not good for the wrench, not good for the metal the wrench is touching, and quite possibly not good for you.

Once the negative terminal's undone, position the cable aside so it can't accidentally remake contact with its spot on the battery. Then undo the positive one.

At this point, you might as well pull the battery out of the plane. There may be some hold-downs to unbolt. When free, the battery should lift right out.

Clean the battery terminals and cable ends with a stiff metal brush; these are available at auto-parts stores.

Take a look inside the battery box. A little wet in there? Dump some baking soda inside and flush it out with fresh water. Wipe it down with soda-sprinkled paper towels.

Lift the battery back into position. Wipe Vaseline liberally over the terminals and the aircraft's cable lugs. It doesn't interfere with power transfer, but it does slow corrosion.

Connect up the positive terminal. When that's on and tight, connect the negative terminal.

Close up the battery box, and you're done.

Bulb replacement

According to Part 43, the owner is allowed to replace ". . . bulbs, reflectors, and lenses of position and landing lights."

Easy enough. Usually it's just a few minutes work to remove the lens or cover with a screwdriver (Fig. 12-33), swap the old bulb for a new one, and replace the cover. A couple of gotchas to watch out for.

First, before putting the new bulb in, take a look at the socket. Sometimes they get wet and a bit rusty. Shoot some WD-40 into it prior to inserting the new bulb. Second, landing lights and some position lights are covered by Plexiglas lenses. Plexiglas stretches and shrinks with temperature changes (well, so does *everything*, but plexi does it a lot more than aluminum). Since the lenses bolt into aluminum housings, the screw holes in the glass are a bit larger than they have to be. It lets the Plexiglas creep back and forth with temperature changes.

Fig. 12-33. *It's easy to replace landing and position lights. However, over-tightening of the screws can cause the Plexiglas covers to crack.*

If you tighten the attachment screws too much, it won't let the Plexiglas creep. A few weeks or months later, you'll find cracks running between the mounting holes. So don't tighten them all the way. The screws generally go into self-locking anchor nuts, so they aren't dependent upon mounting torque to hold them in place. Leave the Plexiglas a tad loose.

Finally, reports indicate that the orientation of landing light bulb filaments can have an effect on the bulb's longevity. When possible, orient the filament vertically instead of horizontally. If bulbs still tend to burn out quickly, experiment with various orientations.

FOR FURTHER INFORMATION

There are several sources of detailed information on aircraft systems and maintenance. McGraw-Hill's *Aircraft Systems* by David A. Lombardo is an excellent reference, and belongs on every aircraft-owner's bookshelf.

Belvoir Publications produces two magazines of interest to the aircraft owner. The first, *Aviation Consumer*, has been mentioned previously. *Aviation Consumer* publishes a lot of mechanical information of use to any pilot.

Another of their magazines is aimed at both professional mechanics, and those owners seriously interested in maintaining their planes. *Light Plane Maintenance* (LPM) (*www.lightplane-maintenance.com*) provides in-depth articles on aircraft operations and maintenance issues. The pricey LPM takes an aggressive stance and is entirely supported by subscribers (i.e., no advertisements).

For subscriptions to either magazine, write to:

Belvoir Publications, Inc.

P.O. Box 420234

Palm Coast, FL 32142

(800) 829-9085

Information published in LPM has been compiled in *Lightplane Maintenance: Aircraft Engine Operating Guide*, by Kas Thomas. It is aimed at pilots of all skill and experience levels, and addresses problems and procedures applicable to all piston aircraft engines.

A highly-recommended reference for those serious about engine operation is the *Sky Ranch Engineering Manual*, by John Schwaner. It provides excellent explanations of how failures occur, what the operator can do to minimize them, and what is involved with the repairs.

13

Problems?

It'd be nice if your airplane ownership went smoothly, with nothing to mar the joy of flight. That's the way it goes for most aircraft owners. But there's always that niggling little worry. What if it breaks down at some little field somewhere? How do I get the plane back if I crash-land somewhere (Fig. 13-1)?

In this chapter, we'll discuss some of these potential problems. Some are minor, some aren't.

DEAD BATTERIES

Considering all the other calamities that might befall you, finding a dead battery in your plane is pretty tame.

In the last chapter, we talked about the electrical system and how a faulty master solenoid can drain the battery. Of course, sometimes they're dead for a less-exotic reason: You left the master switch on.

Finding the problem

If the master wasn't left on, there are two possible sources of the problem: the battery or the master solenoid. A minute or two of troubleshooting is called for.

Listen when you turn on the master switch. Do you hear the familiar "clunk" of the solenoid activating? If so, there's a good chance the problem lies in the contacts of the solenoid rather than in the battery. To make sure, turn on the cabin light. If you heard the solenoid activate, but there's not even a dim glow from the bulb, the solenoid is probably bad.

Find a mechanic to install a new one. Yes, the engine can be started by hand-propping. But it's a real, *real*, stupid thing to do if you've never hand-propped before.

With a bad solenoid, hand-propping may get the engine started but won't solve the power problem. Alternators require an external power source to excite their field windings at startup. If the battery is completely flat, your alternator won't work, and

Fig. 13-1. *How are you going to get your plane back to the airport if you land off-field? This mode of transportation is limited to short distances only.*

could be damaged if the engine is started. Generators also require an external source when first started, though they get by on less. However, running without connection to a battery is a very bad idea—they depend on the battery to help regulate the output voltage.

All right, but, what if you don't hear the solenoid activate when the master switch is turned on? Is the battery discharged or is the solenoid bad?

The best solution is to open the battery case and measure the battery voltage. Radio Shack sells several kinds of *multimeters,* which measure voltage, current, and resistance. Resist buying the larger, more expensive ones, unless you can use them in other electrical work. For trouble-shooting aircraft batteries, the cheap models are adequate.

Set your meter to the lowest range that will measure your battery's nominal voltage (i.e., the 0–15 volt or 0–20 volt range for a 12-volt battery). Touch the red multimeter probe to the battery's positive (+) terminal, and the black one to the negative (–) terminal. If the battery is fully charged, it should read at least its rated voltage; any less is cause for suspicion. And if it's apparently OK, leave the red probe on the positive terminal and touch the black probe to a bit of exposed metal structure. If it doesn't read the same voltage, the ground cable from the battery to the aircraft structure has a poor connection.

If the ground is apparently OK, move the red probe to the input part of the solenoid (you might have to peel back a rubber boot), keeping the black probe touching the metal. If you still get voltage, the problem lies in the solenoid. Or, just perhaps, in the master switch itself or its wire to the solenoid. Turn on the switch, hold the red probe to the positive terminal of the battery, and touch the black probe to the small connector on the solenoid. If you read battery voltage, the switch and wire are good and the solenoid needs replacement.

And if the problem is the battery?

Charging a dead battery

The best solution to dead-battery woes is to remove the battery and charge it.

Disconnect the battery as described in the previous chapter. Remove it from the airplane and bring it to the charger. Don't place the battery directly on a metal or concrete surface; they act as heat sinks.

Remove the battery caps. If it's a lead-acid battery, check the fluid level and add distilled water if necessary. Then connect it to the charger—red clamp to the positive (+) terminal, black to the negative (–). Turn on the charger. The meter on the charger should show a fairly high rate of charge. Go away for a few hours. When the charge rate has dropped to a moderately low value, disconnect the battery, replace the caps, and reinstall it in the aircraft.

Jump-starting

If the battery isn't completely dead, but just can't seem to turn the prop over, there are a couple of other options.

Hand-propping can get the engine started, and if the battery isn't completely dead, it may excite the alternator enough to get it working again. If not, even if the engine starts, the battery won't charge and the avionics won't work.

If your plane mounts a Lycoming and you've already tried to start it, hand-propping will be a bear. The starter Bendix stays engaged. When you prop the engine, you'll also be turning the starter through the ring gear and Bendix. Maybe it doesn't add that much more friction, but the grinding sounds horrible.

But whether your engine is a Lycoming or Continental, don't even think of trying to hand-prop it unless you've had training. Instead, jump-start the airplane just like you would a car. In fact, if your plane has a 12-volt electrical system, you can jump it from your car.

Either way, start by removing your plane's battery caps. If you're working alone, make sure the plane is tied down, the chocks are in place, and the parking brake is set.

If the plane has a power receptacle (Fig. 13-2), you'll need a special set of jumper cables. They can cost a bit, but they make jump starting a lot simpler. Hook the cables up to your car's battery and plug in the connector. Without a receptacle, things get a bit hairier. Some planes have their battery in the baggage compartment. That's not so bad, though it can be awkward to get to.

On most, it's forward of the firewall. This gets a little sporty. Once the engine starts, you'll be working pretty close to the prop to get things buttoned up again. Hopefully, you'll be able to get at the battery through some small access door. But if you've got to remove the top cowl to get to the battery, count on running the engine for a while, then shutting down to reinstall the cowl.

Connect the positive cable (red) to your car's battery, then to the positive terminal on your aircraft. Connect the negative terminal (black) to the aircraft's battery, then to a grounded point under your car's hood. Run your car at a higher RPM for a while to put a little power into your aircraft's battery, then go ahead and start the airplane.

Fig. 13-2. *External power jacks make it easier to jump-start an airplane, but you need a cable with a plug that matches it.*

Jump-starting isn't really the best thing for your airplane. The alternator will stay healthy longer if it doesn't have to provide the massive charge that a flat battery demands. Jump-start as little as possible. Moreover, after the engine starts, the cable system has to be disconnected. There's no way to do it yourself unless you shut down the engine first.

STORAGE

There comes a time for some owners when the airplane can't be flown for a while. Maybe you live in an area where bad weather shuts down flying for months on end. Perhaps you've developed a temporary medical condition, which will prevent you from flying for a bit. Or maybe the money just isn't there.

Inactivity is hard on airplanes. Tires stiffen; engines rust internally. You do an airplane no favors by letting it rest. It's gonna deteriorate on you anyway—fly it, if possible. Otherwise, there are a few precautions you can take, depending upon how long the plane's going to be laid up.

Short-term inactivity

If you won't be flying for just a month or so, there are a few things to do.

First, fill the fuel tanks. Rubber fuel tank bladders dislike exposure to air. Use avgas, as auto fuel is more prone to breaking down over time.

Remove the battery and take it home. Put it on a trickle charger one night a week.

Put some desiccant in the cabin to reduce dampness. Hardware stores carry products to reduce humidity in storage areas. One brand is "Dry-Z-Air."

Cover all ports and inlets to keep the bugs, birds, and rodents out. If there are lots of mice in your area, consider putting metal shields around your tires to keep them from climbing the landing gear into the wings or cabin. Set some traps inside and outside.

Spray exposed hinges and linkages with a heavy corrosion-preventative oil. Your mechanic can probably recommend a particular brand. I use one called "LPS-3." Spray brake disks with light oil; remember to wipe it off before the next flight. Wipe preservative on all exposed rubber parts.

Every week or two, take your battery out to the airport and hook it up. Turn on the strobe and your radios. Tune your radio to all its frequencies and turn the volume back and forth to wipe off any corrosion on the contacts and wipers. Warm the radios for at least 15 minutes.

Roll the plane forward and back a few times to exercise the bearings. Don't put it back in exactly the same position; set it slightly differently so that another part of the tire is touching the ground. If all else fails, bring your jack, lift each tire a bit, and spin them for a minute or so.

Turn on the fuel valve and exercise it to all positions. If you've got a boost pump, switch it on for a second. Rubber parts in the carburetor can tend to dry out, too.

Work the throttle, mixture, and carb heat controls back and forth. Lower and raise the flaps. Move the control wheel/stick gently to its limits a few times. Check the preservative oil you've sprayed on the hinges and renew if necessary.

Making sure the magneto switch is off, the chocks are set, and the tiedowns are tight, step to the front of the airplane and turn the engine over by hand. Do at least ten blades (five complete turns of the engine).

Do *not* start the engine and idle it. Combustion byproducts include water vapor, which reacts with the other elements to form acids and other nasty stuff. Collection of these harmful materials is halted by getting the engine warm enough to drive the moisture out. The engine won't get warm enough during a ground runup.

Turning it over by hand wipes any condensed moisture off the cylinder walls, lowing the advance of corrosion—do this as often as possible. Engine manufacturers generally recommend turning an inactive engine by hand at least once a week.

Your plane should be able to weather a month or two using these procedures; more, perhaps, during cold weather. It's not *good* for it; but no serious harm will likely occur. If you can still fly it occasionally, the storage period can stretch out.

To be sure, get advice from your mechanic.

Long-duration storage

Not going to be able to fly your plane for six months or a year?

Sell it.

Long-duration storage is hard on your airplane. It may never make it back to the level of reliability it had. Corrosion takes its toll, both on the airframe and inside the engine. Avionics connectors develop internal oxidation, and strange problems result.

Grease runs out of wheel bearings, control pulleys, and propeller hubs. You're better off, in the long run, to cut your losses and sell. Buy another plane when you're able to fly again.

Oh, all right. It's not always an option. Sometimes the plane fits your needs perfectly; you're loath to get rid of it. Maybe it's a rare model. And admit it: sometimes we get emotionally attached to our metal birds.

Get with your mechanic and plot out the storage plan. He or she will probably recommend the following course:

1. Draining all the engine oil and replacing it with preservative oil
2. Spraying the inside of the cylinders with hot preservative oil
3. Replacing the spark plugs with special plugs containing desiccant (Fig. 13-3)
4. Blocking the exhaust pipes with desiccant plugs
5. Spraying the inside of the airframe with a corrosion fighter such as ACF-50
6. Putting the airplane up on blocks to preserve the tires
7. Taping/sealing all gaps

Storage location is important, too. In many parts of the country, outside long-term storage is not practicable. You'll need a roof over its figurative head.

If you've got a good-sized garage, you could take the plane apart and store it there. But if you ever decide to sell it, you'll take a real loss. Flyable airplanes are worth much more than disassembled "projects."

Taking it out of storage is a major undertaking as well. In all probability, it'll be due for an annual, anyway. But a lot of bad things can happen if a plane sits inactive for too long. Brace yourself for a steep bill.

Fig. 13-3. *Desiccant plugs are clear plastic and contain crystals, which absorb moisture from within the engine. The crystals change color to indicate when they must be replaced.*

PROBLEMS AWAY FROM HOME

Few things are as stressful as having your plane break down on a trip (Fig. 13-4).

Tools

Consider stowing a few tools in the baggage compartment. Nothing heavy or large; just some stuff that might just get you going if something minor happens. This assumes you're used to maintaining your own plane; don't try to learn on a windswept ramp 200 miles from home.

You can usually get by with a small tool box, or perhaps a cloth bag. Put a 3/8-inch drive socket set in it (hardware stores sell cheap small ones suitable for occasional work), some screwdrivers (both types of tips), a set of common wrenches (3/8 inch, 1/2 inch, 9/16 inch, 5/8 inch, and 3/4 inch), a couple of types of pliers, some safety wire, wire twisters, and others. Add a sprinkling of extra nuts, washers, and cotter pins.

Keep your spark-plug socket in the box, plus some extra copper washers. Save a couple of old plugs from the first time you install new ones. Wrap them carefully, stuff them inside a small cardboard box or toilet-tissue tube, and tape it shut.

Fig. 13-4. *Stan Brown contemplates a solenoid failure on Ramp Rooster (see Chap. 6).*

Same thing if you ever replace an inner tube—compress an old one down and store it in the far corner of the baggage compartment, especially if your plane uses an unusual tube. I got a flat tire at a remote grass strip once (Fig. 13-5). I took the wheel off (using willing bystanders as a jack) and had all the tools to separate the wheel. But no one on the field had the right size tube.

Keep the number of onboard tools within reason, of course. Most of the time, fixing it yourself is just wishful thinking.

Finding help

If the problem is beyond the normal run of "preventative maintenance" items (for example, bad plug, flat tire,and the like), an A&P will be necessary. At most airports, they should be easy enough to find during working hours on a weekday. They're usually prepared to slip transient aircraft into their schedules, to at least diagnose the problem.

On weekends, most FBOs and flight schools have one on call. Unfortunately, the on-call rate is higher—you may end up shelling out $120 or more an hour.

Free-lance A&Ps might be a better pick. Wander around the airport and talk to some owners. Explain your problem and ask for an A&P recommendation. Free-lancers often work on weekends.

Ferry permits

Sometimes you'll have a problem that doesn't make an airplane physically unflyable, yet is not legally airworthy. A problem with the gear retraction system, for instance. Or you get weathered-in at a remote airstrip and your annual expires.

Fig. 13-5. *The Fly Baby's left tire went flat at this remote strip during a picnic. There were plenty of people to lift the airplane and tools to remove the wheel, but no replacement tube. The plane was secured, and the author caught a ride home to buy a new tube.*

Since the airplane does not then correspond to the requirements of its type certificate, its airworthiness certificate is invalid until the fault is corrected.

The FAA can authorize flight under these conditions using a "special flight permit," commonly known as a *ferry permit*. They're covered under 14CFR 21.197:

"(A) A special flight permit may be issued for an aircraft that may not currently meet applicable airworthiness requirements but is capable of safe flight, for the following purposes:

1. Flying the aircraft to a base where repairs, alterations, or maintenance can be performed, or to a point of salvage."

A special flight permit actually is a temporary replacement for your aircraft's airworthiness certificate. In fact, you'll use FAA Form 8130-6, which is the application for an airworthiness certificate. You can find it at FAA offices, or can download the form. Do an Internet search for "FAA Form 8130-6."

Apply via an FAA Flight Standards District Office either in person or by phone. Have all your paperwork ready and be prepared to explain the situation to the inspector.

If he or she approves your application, you'll receive a temporary airworthiness certificate. If you've applied by phone, the FAA will fax it to you. This form must be kept in the aircraft.

Prior to the flight, an A&P must examine the aircraft and certify that the plane is safe for the proposed flight. The FAA may require certain additional steps be taken. For instance, say your retractable aircraft has a landing gear problem. The FAA will authorize you to fly the aircraft with the gear locked down. They will also require that the pilot be unable to retract the gear. The A&P will have to disconnect the switch or take some other means to prevent retraction.

A couple of things to be aware of. First, the special flight permit exists to allow flight to a location where maintenance can be performed, not to let you fly to a place where you can get it done cheaper! If there are extenuating circumstances, such as wishing to fly the plane to a specialist in your particular aircraft, bring them to the attention of the inspector. Second, the permit only authorizes solo flight. If you're on a family jaunt, they'll have to get home another way. Third, the permit authorizes one flight. Once you reach the destination, the permit expires. Finally, check your insurance. It may not cover your aircraft when operated under a special flight permit. However, ferry permits can also be granted for "Evacuating aircraft from areas of impending danger." If you're based on the Gulf Coast with a crippled airplane and a hurricane is coming, you can receive FAA permission to fly the plane out of the danger area. Contact your insurer in these cases; they may cover the evacuation flight.

OVERHAULS

"You need an overhaul."

Ugh. That's probably the worst thing you can say as an airplane owner.

I can't give you many words of comfort here. There's no question that it's going to cost big bucks. Not that don't have any options. Just no low-cost options.

When?

When do you overhaul the engine? As has been discussed previously, the engine's TBO has no legal significance in noncommercial operation. As a private operator, you can legally fly it until it comes apart.

However, it's best to overhaul *before* the engine comes apart.

There are a number of "triggers," which commonly initiate the overhaul process. The first, of course, is catastrophic failure. A connecting rod jutting from the crankcase is Mr. Lycoming's way of saying, "Get out the checkbook." It doesn't have to be quite that spectacular—the sudden appearance of large pieces of metal in the oil is another sign.

Fortunately, sudden catastrophic failures without warning are rare. In most cases, warning signs just aren't picked up. Detecting them might save you a forced landing, but won't necessary delay the inevitable overhaul. The best way to gain some warning is through oil analysis—send a sample to a testing agency at every oil change. They'll track the trends, and will warn if there's a sudden increase in wear metals in the oil.

Another overhaul trigger is a prop strike. Since the engine manufacturer probably calls for a complete teardown and inspection anyway, you might as well spring for an overhaul.

But probably the most common trigger for an overhaul is just plain lack of power. Over years of operation, springs get weaker, bearings wear down, and tight fits become sloppy. The engine just gets tired.

Cylinder compression is one of the primary indicators of engine condition. Every annual, your mechanic uses compressed air and pressure gauges to measure how "leaky" each cylinder is. As Fig. 13-6 shows, air regulated at 80 psi is applied to a calibrated 1/4-inch long 0.040-inch diameter orifice, with the outlet connected to the cylinder via a spark-plug hole. The air pressure is measured on each side of the orifice. The cylinder's piston is at Top-Dead-Center (TDC) of the compression stroke; piston all the way to the top, with both intake and exhaust valves closed.

If there's absolutely no leakage in the cylinder, the pressure gauge on the cylinder side of the orifice indicates 80 psi. But if the piston rings don't quite seal right, or either valve isn't quite seating, some air will leak out. The compression of that cylinder is the ratio of the input pressure versus the pressure inside the cylinder: "80/65" means that 80 psi was applied to the upstream side of the orifice, but enough air leaked through from the cylinder to reduce the pressure by 15 psi.

Where that 15 psi went is informative. Your A&P will use his or her ears. If the carburetor's hissing, the problem is probably with the intake valve. At the exhaust pipe? Exhaust valve, of course. If the hissing comes from the crankcase breather, the rings are most likely the culprit.

A cylinder with 80/65 compression is on the low side, but acceptable. It could fly for years at this level. Compression will vary both up and down, depending upon a number of factors. Normally, though, you'll see a gradual trend downward.

Compression lower than 80/60 usually grounds the engine. If one cylinder is low, but the others are acceptable, you may be able to get by with just repairing the offending cylinder. Sometimes, the only problem is a little carbon around the valves.

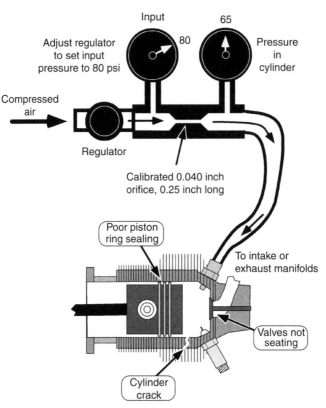

Fig. 13-6. *Cylinder compression testing.*

If they're all low and the engine is still far from its nominal TBO time, a "top overhaul" (Fig. 13-7) is a cheaper alternative. In a top overhaul, the cylinders, valves, and pistons receive the full treatment but the crankcase (the "bottom") is left undisturbed. The engine doesn't even have to be removed from the aircraft. But if the engine has a lot of time on it, you might as well bite the bullet and get a full overhaul.

Finally, be aware that compression checks aren't perfect. A little crud in that 0.040-inch orifice or an improperly-set regulator will make your cylinders read artificially low. If the cylinder isn't right at TDC, one or the other valve might be slightly open. And each piston ring includes a little gap; these gaps rotate slowly around the piston as the engine runs. These gaps can sometimes line up and cause a temporary compression loss.

Don't get railroaded into an early overhaul! Take the plane to another mechanic for a second opinion.

Who's going to do it?

Probably the biggest factor in how much an overhaul costs is in who performs it. Let's look at some options, from cheapest to most expensive:

Fig. 13-7. *Top overhaul in progress. The engine can remain in the airplane while the cylinder heads, cylinders, and pistons are removed for replacement/refurbishment.*

DOING IT YOURSELF

Like any other aircraft maintenance item, you can overhaul your own engine, as long as a licensed A&P supervises and signs off your work.

If you've got an extensive mechanical background and an A&P willing to take the risk on you, this might be the option for you. The smaller aircraft engines aren't that complex; there isn't much more to a Continental A-65 than to a VW engine. If you can hot-rod a Chevy V-8, a Lycoming O-235 should hold no surprises. You'll be left with a less-lightened pocketbook, plus a full understanding of the engine's condition.

Unfortunately, it's not the solution for most of us. Too much can go wrong; unless you've got the right background it'd be best to leave the engine to a pro.

FREELANCE MECHANIC

The next step up is to find an A&P willing to overhaul the engine for you. This is probably the lowest-cost realistic solution to your problem.

Unfortunately, it, too, is filled with risk. How good is the mechanic? Will they cut corners to maximize their profit? After all, your only way to judge is by how good the engine looks afterward. Paint is a lot cheaper than gaskets, seals, and pistons.

My friend Kirk hired a freelancer to overhaul the engine for his homebuilt. After completion of the overhaul, Kirk damaged the engine during installation and took it to another shop for repair. The new shop refused to sign off the engine. Many parts were beyond tolerance; the interior of the engine was filthy. Kirk paid $3000 more to get the engine into legal condition.

Remember, too, our case study in Chap. 3: The initial rebuilder held the engine for a year and still showed no sign of progress.

Still, a freelancer is the cheapest route. If you've got a hot prospect, ask for references. Talk to those people who are flying engines the A&P overhauled. See how they're holding up.

FBO OVERHAUL

Overhauls performed by the local FBO will cost more, since they've got more overhead. Yet the quality can vary, just like a freelancer. Nominally, at least, their shop may be better equipped than someone working in their garage. Plus, you may prefer to deal with a business rather than a private individual. But FBO overhauls can also suffer from a desire to maximize the bottom line.

REPAIR STATION OVERHAULS

Some companies have specialized in engine overhauls. These outfits stake their reputation on their work; hence you'll see a uniformly high level of quality.

You'll usually see them listed as "certified repair stations." These are governed by Part 145, which places stringent requirements on facilities, equipment, personnel, and recordkeeping. Companies can gain certification in a number of fields, such as engines, instruments, propellers, or airframes.

The FAA keeps a tight rein on repair stations. They're certified to perform work on rather limited aspects of aviation, such as "reciprocating engines of 400 HP or less," or "instruments/mechanical." Stations can gain ratings in more than one category, of course.

You can expect the highest-quality work from a certified repair station, as well as the highest prices. Yet it can pay off, when time comes to sell the airplane. Listing the engine as "400 SMOH by Victor" has a certain attraction.

One strange aspect of having an overhaul done by a certified repair station—your engine probably won't be overhauled by an A&P! The A&P rating requires expertise in a wide range of skills, from engines to fabric to riveting. Employees of repair stations gain "repairman certificates," which allow them to perform specified functions while under the station's employ. And since they can specialize in one area rather than many, they can become very good at it.

FACTORY OVERHAULS

The engine manufacturers have gotten into the act. Both "service" and "new" limit overhauls are available, depending upon the engine type.

A new limit overhaul by the original builder of the engine is called a *remanufacture*. That, again, has a certain cachet. A certified repair station can perform the same sort of overhaul, but is not legally authorized to call it a remanufacture.

Factory overhauls feature their own little bizarre twist. By 14CFR Part 43.3, the factory overhaul doesn't have to be performed by a person with an A&P or repairman certificate, and the factory itself isn't required to be a certified repair station.

Other aspects

Many engine accessories also require overhauling at particular intervals. Depending upon each accessory's status, you may choose to have them overhauled at the same time as the engine. If you can afford it, go ahead—it'll probably be a lot less trouble in the long run.

Many overhaulers don't actually overhaul your engine for you. Rather, they give you an overhauled engine in exchange for money and your old engine, called a *core*. If your old engine has significant flaws (cracked case, bent/broken crankshaft) there will be additional charges.

Also, if you buy an overhauled engine from nonlocal company, you'll pay more than just the overhaul cost. A mechanic will have to remove your old engine and crate it. Shipping charges will run several hundred dollars, if not more. When the new engine arrives, the mechanic has to install it and check it out (Fig. 13-8).

This process may take a while. You'll probably want to treat the airframe for storage. If the plane's equipped with tricycle gear, taking the engine out will shift the CG back so far that the plane will want to sit on its tail. You'll either need to put something under the back end or, like the owner of the plane in Fig. 13-9, put some heavy ballast in the engine compartment.

Since the major overhaulers swap a rebuilt engine for your core, your plane might not have to stagnate. Some companies offer complete engine replacement services. If the engine is still safely flyable, you can fly into the site, hand over the plane, and pick it up a day or so later with the brand-new engine already installed.

Alternatives

Rather than facing an overhaul on your engine, there are a couple of other routes you could take.

What about a used engine? They turn up on the markets occasionally. The Florida hurricanes of 2004 destroyed a lot of airframes; yet the engines were generally intact.

Fig. 13-8. *Installing the engine on an Aeronca Champ.*

Fig. 13-9. *This block of concrete holds the plane in normal ground attitude until the overhauled engine is installed.*

The insurance companies (or owners, for those planes without hull coverage) put them up for sale.

There are one or two drawbacks to buying a used engine. First, the engine may have been inactive for quite a while. You'll end up paying for your A&P to verify the condition of the engine. Second, it may be difficult to find the *exact* model of engine you need on the used market. If your aircraft uses a Lycoming O-320H2AD, then you can't install just any O-320. It must be either an O-320H2AD, or you must go through the STC process to gain approval.

Finally, the competition's stiff. The O-320s used in Cessna 172s, for instance, are rabidly sought by those building RV-4 and RV-6 homebuilts.

If you're less than enamored by the performance of your aircraft, overhaul time is the perfect opportunity to solve your problem. If you're going through all the work to remove, overhaul, and replace an engine, why not upgrade to a bigger one? A number of commercial STCs are available to allow installation of larger engines on the common aircraft of the General Aviation fleet.

INCIDENTS AND ACCIDENTS

"Stuff happens."

What do you do if it happens to *you?* No one likes the thought of an accident or forced landing. Let's take a few moments and consider some of the factors.

Primary consideration

The primary thing you should be worried about is medical attention for any injured parties. All else pales in comparison.

Reporting requirements

Aircraft accident investigation is handled by the National Transportation Safety Board (NTSB). The FAA will get involved, but the NTSB is in charge. Accident reporting requirements are contained in the NTSB's Part 830.

If the aircraft receives substantial damage, or any person suffers serious injury or dies, the event must be reported to the NTSB "by the most expeditious means possible."

As ever, definitions are important. *Serious injury* is defined as broken bones other than fingers, toes, and noses (ouch!), severe bleeding, nerve, muscle, or tendon damage, internal organ damage, second- or third-degree burns, or any injury that requires hospitalization for more than 48 hours.

The NTSB defines *substantial damage* to include damage or failure, which affects the structural strength, performance, or flight characteristics of the aircraft, and would require major repair or replacement (Fig. 13-10).

What's more interesting is what the NTSB does *not* consider substantial damage: engine failures, bent fairings or cowlings, dented skin, small punctures in skin or fabric, ground damage to a prop or rotor blades, and damage to landing gear, wheels, tires, flaps, engine accessories, brakes, or wingtips. These sorts of events qualify as *incidents*. The NTSB defines this term as "An occurrence other than an accident associated with the operation of an aircraft, which affects or could affect the safety of operations."

Why is this important?

Simple. Aircraft accidents are reportable. Incidents, except for certain exceptions, aren't.

If you ding a wingtip on a post while taxiing and fracture the fiberglass, you don't have to tell the NTSB or the FAA.

If your engine quits and you safely land in a field, the NTSB doesn't want to know about it (Fig. 13-11).

Fig. 13-10. *This accident probably meets the NTSB's definition of "major damage," and thus must be reported.*

Fig. 13-11. *This Cessna 175 set down in a farmer's field and went through a fence when a clogged fuel line starved the engine. The damage to the cowling, nosewheel pant, and spinner are not sufficient to qualify as "substantial damage" by NTSB criteria. It is thus an "incident," and need not be reported.*

The exceptions are pretty limited. You must inform the NTSB of any incident involving flight control system malfunction or failure. Any in-flight fire, any midair collision (just consider yourself lucky if you're *around* to file the NTSB report), and any incident where the loss of property (other than the aircraft) exceeds $25,000, and, again, if anyone suffers serious injury or death.

Of course, you'll have to report any accidents (and probably most incidents) to your insurance company.

Securing the aircraft

If you've had an accident, the NTSB requires that you preserve the wreckage. You can't start hauling pieces away, and the general public should be prevented from doing the same. Report the accident to your insurance company as soon as possible; they should handle the security arrangements.

If you've set down in a field somewhere, use your tiedown kit to secure the plane until it can be recovered.

Recovery

Whether you've had an accident or just force-landed off-field, the plane needs to be removed from the scene. How you get it out depends on a number of factors.

Is the plane intact and flyable? You could fly it out, if the property owner doesn't mind. If you landed on a road, you're probably out of luck. Many areas have ordinances prohibiting take-offs from public property, including roads.

Are your *skills* up to it? It's one thing to do soft-field practice on an asphalt runway. It's another thing entirely to try to drag your 172 out of a fallow field with trees in the distance.

Finally, your insurance probably prohibits it. Take a look at your policy—it probably allows operation only from approved airfields.

If you have an in-flight hull policy, it should include a clause covering "transportation to the nearest airport in the event of an off-airport landing." Secure the airplane and contact your insurance company.

If you don't have hull? You might have a chance. Your liability coverage may apply. Check your policy or talk to an agent.

Similar factors hold if the aircraft isn't flyable. If the plane is totaled, the insurance company will take possession. If not, your hull coverage should handle transportation.

The means of transportation of your aircraft depends upon its proximity to roads. Hauling it from an isolated mountain meadow will probably require a helicopter. Sometimes the plane can be carried intact; otherwise it'll have to be partially disassembled in place. Just leaving it there may not be an option, either. All the more reason to carry hull coverage.

If the site is accessible by road, recovery will cost quite a bit less. Again, if your insurance covers it, leave it to them. Otherwise, you'll have to either hire a specialist or bring it home yourself. Typically, a large trailer is pulled to the landing site (Fig. 13-12). The fuel tanks are drained (assuming they weren't empty already. . . .) and the wings removed. The fuselage is pulled onto the trailer using either manpower or a winch. If the horizontal stabilizer is wider than highway laws allow, it

Fig. 13-12. *Aircraft recovery trailer. In this case, it's being used to tow the airplane in a parade.*

must be removed, too. External locks are applied to the rudder and elevator (if still attached).

The wings are either stowed on the trailer or carried separately. They must be secured during transport, or they might well blow off the trailer. If wooden supports are to be used, they'll either have to be built back at the airport or cut by hand at the landing site or bring a portable generator to run your circular saw. Plenty of padding will be necessary if you want to protect the paint job.

You can disassemble the plane and place it on the trailer, but an A&P will have to supervise and sign off the reassembly (Fig. 13-13).

Fig. 13-13. *Cessna wings are easily removable for transport, but their reinstallation must be supervised and signed off by an A&P.*

14

The Answer

When you bought this book, you probably had one overriding question—How much does it cost to own a plane?

Ask owners, and some will grin and say, "Everything you have, plus ten percent."

And if you press for actual numbers, they may not be able to give you an accurate estimate. I talked to a lot of owners while researching this book. Some weren't sure how much their metal steeds drained their pocketbooks. "I don't *want* to know," said one man. One of the case study participants gave me several pages detailing his expenses over a number of years. At the end was the admonition, "Don't tell my wife!"

Another common thread appeared during the interviews. Some commented, "I spent a lot on parts the first year, until I found this place. . . ." or, "I really got took during my first annual, but a guy told me about this other outfit. . . ."

Like flying itself, there is a learning curve to airplane ownership: The longer you've had a plane, the more connections you make and the cheaper it becomes.

So what should you do?

Simple: Buy an airplane.

Start looking for a good used 152, 172, or Cherokee. Get a good prepurchase inspection. Keep some money in reserve in case of postpurchase problems.

Fly it for a year, then take a look at your finances. Are you affording it? Is your cash reserve gone? Is the plane just too much of a drag on the family budget?

Then sell it. You'll probably make money. If you've only flown a hundred hours or so and haven't hurt the airplane, you'll have little trouble finding a buyer. Prices are skyrocketing; I sold my 150 for $5500, and 20 years later, they're getting three times that for good used 1965 models.

Be warned, though. There's a danger—once you have owned an airplane, you'll never be able to rent again.

Sure, you can call the local FBO, and schedule yourself some 172 time. But after your 1.3-hour ("Return the aircraft ten minutes early for the next renter") flight, you'll slowly carry the keys back to the office. You'll listen to the "beep-beep-beep" of the credit card machine with the sky still singing in your blood.

You'll think about those old biplanes flocking around that little strip the FBO won't let anyone land at. You'll remember that family picnicking under the wing of their 172. The icky-yet-pleasant labor of scrubbing the bugs off the leading edge, watching the gleam of your very own airplane emerge from beneath the suds and bug parts.

You'll be ruined, all right.

But what a way to go. . . .

GLOSSARY

100 LL   One-hundred octane low-lead aviation fuel.

100-hour inspection   Similar to an annual inspection; performed by an A&P.

A&P   Person with an Airframe and Powerplant rating; authorized to perform most inspection and maintenance on aircraft.

Accessories   Officialese for engine-driven items not built by engine manufacturer. Includes magnetos, starters, alternators, and so forth.

Accessory case   Portion of the engine, which contains gearing to turn aircraft accessories such as magnetos and alternator.

ACO   Aircraft Certification Office.

AD (Airworthiness Directive)   An FAA announcement mandating certain inspections or repairs of given aircraft or equipment.

Airknocker   Nickname for the Aeronca Champ.

AN   Air Force-Navy standard for aviation hardware.

Annual inspection   FAA-required in-depth inspection and analysis of aircraft condition; must be performed by an A&P with an IA authorization.

Antique   An aircraft built prior to WWII.

AOPA   Aircraft Owners and Pilots Association.

Autogas STC   Aircraft has FAA approval to run on automobile gas.

Autogas   Automobile gasoline.

Avionics   Aircraft electronics, typically for communications or navigation.

Basket case   Slang term for a disassembled aircraft that needs restoration from the ground up.

Chocks   Devices to prevent aircraft from rolling when parked.

Cigarette   Slang term for ceramic insulator at the end of aircraft spark plug wire.

Classic   An aircraft built from 1945 to 1955. A "contemporary classic" was built from 1956 to 1962.

Communal hangar   Large building storing multiple aircraft with a single door for aircraft entry/exit. Planes often have to be repositioned to allow others ingress or egress.

Conventional gear   Landing gear configuration where the main wheels are mounted well forward and a smaller one is under the rudder.

CS or C/S   Constant Speed (propeller).

DAR   Designated Airworthiness Representative.

DER   Designated Engineering Representative.

Direct costs   Costs associated directly with flight; that is, fuel and oil expenses.

EAA   Experimental Aircraft Association.

ELSA Experimental Light Sport Aircraft. It is a homebuilt aircraft based on a certified Special Light Sport Aircraft design.

ELT Emergency Locator Transmitter.

Endless loop See recursion.

Ethanol A form of alcohol which cannot be present in automobile fuel used in aircraft approved for its use.

Experimental A certification category covering a number of subcategories. Usually refers to "amateur-built" (homebuilt) aircraft.

FAA Federal Aviation Administration.

FBO Fixed Base Operator.

FCC Federal Communications Commission.

Ferry permit FAA authorization to fly an aircraft, which does not currently meet airworthiness requirements to a location where it can be repaired.

Fixed costs Those costs which are connected with the ownership, and not operation, of the aircraft. It includes expenses like hangar rents, insurance premiums, annual inspections, and the like.

Flat-rate annual Annual inspection performed for a given price. Price may actually be higher if necessary repairs exceed a certain level.

FSDO FAA Flight Standards District Office.

GPS (Global Positioning System) Satellite-based navigation system.

Groundloop Rapid ground maneuver performed by out-of-control taildraggers. Occasionally damaging; seldom fatal, always embarrassing.

Handprop Starting an aircraft engine by manually turning the propeller.

Homebuilt Aircraft built by a private individual for education or recreation.

Hull coverage Insurance coverage for actual physical damage to the aircraft.

IA A&P with an Inspection Authorization; allowed to approve modifications and major repairs, in addition to performing annual inspections.

Impulse coupling Portion of magneto that alters ignition timing to aid starting.

JAFCO "Just another (Friendly) Cessna One-Seventy-Two" (or One-Fifty, and so on).

Kitplane Homebuilt constructed from a kit rather than from raw materials.

Leaseback Allowing your aircraft to be rented out in exchange for a given hourly payment.

Liability coverage Insurance that covers damage to passengers or third parties in the event of an accident or incident.

Life-limited Aircraft part which must be replaced on a given interval; hydraulic hoses which deteriorate over time are an example.

LPM Light Plane Maintenance magazine.

LSA Light Sport Aircraft, an aircraft which, having met certain characteristic and performance requirements, can be legally flown by persons executing Sport Pilot privileges.

MDH Major Damage History.

Milkstool Taildragger pilots' derogatory term for tricycle-gear aircraft.

Mogas Automobile gasoline.

MTBE   Methyl Tertiary Butyl Ether, an automobile gas additive which is acceptable in fuel used by aircraft approved to use auto gas.

Multimeter   Small electrical tool used to measure voltage or electrical resistance.

Nav-com   Radio with both navigation and two-way communications capability.

Not-in-flight   Insurance which covers damage to the aircraft as long as the damage occurred while the aircraft wasn't flying (i.e., vandalism, weather damage, and the like).

NTSB   National Transportation Safety Board.

OEM   Original Equipment Manufacture. Replacement parts made by the same company that provided the original component.

Oil screen   Coarse metal mesh, which filters engine oil.

Overhaul   To restore an engine to a condition where it should last its listed TBO.

PMA   Parts Manufacturer Approval; replacement parts are approved by FAA to replace parts originally manufactured by another company.

Prop strike   An incident where the propeller struck something while the engine was turning (see sudden stoppage).

Ragwing   Slang term for fabric-covered aircraft.

Recursion   See endless loop.

Repairman Certificate   Authorization to perform a given set of maintenance tasks. A person who constructed a homebuilt aircraft receives a repairman certificate which allows him or her to perform all maintenance and inspections on the aircraft.

Rotax   A brand of two- and four-stroke engines popular on ultralights and small homebuilts.

Run out   An engine which has reached its TBO.

SDR   Service Difficulty Reports; These are reports of unusual maintenance findings filed with the FAA and regularly disseminated to the aviation community.

Service bulletin   An airframe and/or engine manufacturer's announcement that certain repairs or inspections should be performed to forestall certain failures. Similar to an AD, a service bulletin does not have the force of law.

Service manual   Official publication by airframe or engine manufacturer, which specifies maintenance procedures.

SBS (side by side)   Aircraft seating arrangement where pilot and passenger sit next to each other, rubbing elbows.

Signature loan   A no-collateral loan based solely on the applicant's credit rating.

SLSA   Special Light Sport Aircraft; It is a production aircraft certified under industry consensus standards.

SMOH   Engine time Since Major Overhaul.

STC   Supplemental Type Certificate; It is an amendment to the original TC, often by a source other than the original manufacturer.

Sudden stoppage   Term for when the propeller strikes something that causes the engine to halt. Induces major internal stresses; engine usually has to be torn down and inspected.

Taildragger   An aircraft with "conventional" gear; that is, two large wheels mounted forward and one smaller one under the rudder.

Tandem   Aircraft seating arrangement where the pilot and passenger sit one behind the other.

TBO   Time Between Overhauls.

TC   Type Certificate.

TDC   Top Dead Center; the position where a piston is as far up a cylinder as it can go.

Tiedown   Unsheltered place where aircraft can be parked for long period of time and secured against wind.

Title search   Process by which the legal owner of an aircraft is determined. It should be performed prior to purchase, to ensure the seller is the legal owner.

Top overhaul   An overhaul performed only the compression section on an aircraft engine (pistons, rings, cylinders, cylinder heads, valves, and others).

Tricycle gear   Landing gear configuration where the aircraft's third wheel is mounted in front, like a child's tricycle.

Trigear   See tricycle gear.

TSO   Technical Standard Order; It is a basic standard that the aircraft parts must meet.

Type Certificate   Document officially defining an aircraft's equipment and configuration. Aircraft must conform to be legally airworthy.

Wanttaja   An aviation writer; pronounced as "Wahn-TIE-ah."

Index

Private pilots, 69
Problems, 274–292
    dead batteries, 274–277
    incidents/accidents, 288–292
    overhaul, 282–288
    storage, 277–279
    on trips, 280–282
Prop strikes, 283
Propeller blast, 174–175
Propeller logs, 149, 156
Propellers:
    cost of, 129
    data sheets for, 86, 88
    examination of, 146
Property losses, 290
Proprietary information, 226
Puckette, Margaret, 19

Q
"Quick Build" kits, 127
Quick-drain valves, 246

R
Radios, 44
    FCC licenses for, 166
    planes without, 178
"Ramp Rooster," 110, 280
Ramps, 181, 182, 193
Range:
    of new aircraft, 58–60
    of used aircraft, 76–82
Recreational pilots, 69
Red tags, 219
Redundancy, 99
Registration application, 93, 163, 164
Registration certificate, 119
Registration number system, 166
Remanufacture, 286
Remote mount filters, 244
"Remove Before Flight" banners, 183
Renting, 1, 3, 5–6
Repair station overhauls, 286
Repairman Certificates, 121, 138,
    140, 286
Repairs, logbook notation of, 148
Replacement cores, 256
Replacement parts, 218–222
    for antique planes, 103–106
    for Classic planes, 113

legal, 219
    owner-manufactured, 220–221
    and purchase negotiation, 156
    standards for, 221–222
    Type Certificate for, 218–219
Resale price, 7
Retractable landing gear, 82
Retraction systems, 25–26
Reverse engineering, 226
Riveting, 136
Rock deflectors, 98
Rockwell, 81
Rope tiedowns, 173
Rotating tires, 259–261
Rotax propellers, 129, 132, 133
Rough grass airports, 161–162
Rubber parts, 195, 257, 278
Rudders, 47
Runway length, 41, 42
Runways, uncontrolled-airports, 178
Rust, 136
RV-6 homebuilt, 9
Ryan, 81

S
Saddle tanks, 199
Safety, 63–64, 194–195
Safety wire, 236, 242, 243, 280
Safety-wire pliers, 236, 237, 242, 280
Sales tax, 64, 165
Saratoga II HP, 38
Sato, Bill, 109
SBS (*see* side-by-side seating
    configuration)
Scheduling, annual-inspection, 215
Scheduling services, 162
Schwaner, John, 273
Screens, fuel, 250, 251
Screwdrivers, 185, 236, 280
Screws, 185, 236
SDRs (*see* Service Difficulty Reports)
Sealants, 195
Seasonal changes in gasoline, 197
Seat configuration, 44
    and liability premiums, 167
    of new aircraft, 58–60
    of used aircraft, 76–82
Seating capacity, 42, 43
Secured loans, 85

Wood, 135, 136
Wood-and-fabric aircraft, 120
Wooden construction, 125, 126
"Working rating" (of rope), 173
Workload, 32
Workshops, 135
Wreckage, preserving, 290
Wrenches, 236, 280

Y
Year of aircraft, 76–82
Yellow tags, 149, 219
Yoke locks, 180
Young Eagles program, 117

Z
Zero-timed engines, 92

ABOUT THE AUTHOR

Ronald J. Wanttaja is an award-winning aviation writer and a former systems engineer with Boeing, who worked in satellite orbit/constellation design and analysis, launch vehicle and onboard propulsion system trades, and operations concepts for space systems. He worked on the early design studies for the International Space Station. An Air Force veteran, he was an on-duty operator for the Defense Support Program missile early-warning satellite. As a freelance aviation journalist, he has written for Private Pilot, Flying, Sport Aviation, Flight Line, Kitplanes, and other publications. He is author of the book Kit Airplane Construction, Third EditionTh3/e, also from McGraw-Hill. His aviation writing has won several prizes, including Flying Magazine's "Bax Seat Trophy" and a journalism award from the Aviation/Space Writer's Association. Mr. Wanttaja has also written and published historical fiction.